Edward Step

By the Deep Sea

A Popular Introduction to the Wild Life of the British Shores

Edward Step

By the Deep Sea
A Popular Introduction to the Wild Life of the British Shores

ISBN/EAN: 9783744675482

Printed in Europe, USA, Canada, Australia, Japan

Cover: Foto ©berggeist007 / pixelio.de

More available books at **www.hansebooks.com**

By the Deep Sea

A POPULAR INTRODUCTION TO THE

WILD LIFE OF THE BRITISH SHORES

BY

EDWARD STEP, F.L.S.

AUTHOR OF "WAYSIDE AND WOODLAND BLOSSOMS," "BY VOCAL WOODS AND
WATERS," "BY SEASHORE, WOOD, AND MOORLAND," ETC.

*WITH 122 ILLUSTRATIONS BY P. H. GOSSE, W. A. PEARCE,
AND MABEL STEP*

LONDON

JARROLD & SONS, 10 & 11, WARWICK LANE, E.C.

1896

" There is a rapture on the lonely shore,
   There is society where none intrudes
   By the deep Sea, and music in its roar:
   I love not Man the less, but Nature more."

Byron's " Childe Harold," Canto iv.

CONTENTS.

ILLUSTRATIONS.

ILLUSTRATIONS.

From photo]   THE SANDY SHORE AT LOW-WATER.   [by the Author.

BY THE DEEP SEA.

CHAPTER I.

THE SEA AND ITS SHORES.

THE sea is the very fountain and reservoir of the life of this globe. As the heart is to man and his fellow vertebrates, so is the ocean to the world. It is the centre of the circulatory system; and that system means the life, the health, the sustenance of the body through which it sends its fluids. With the destruction of the heart the human life must cease; and with the annihilation of the sea, could such a thing be possible, all life on the globe must come to an end. We know it is the source of all our vitalizing showers, of every fertilizing stream, of every commerce-laden river. The sun and the winds distil its waters, and carry the sponge-like clouds over the lands, to drop their moisture in rain and mist and snow, making vegetation possible, and giving man two-thirds of his entire substance; for there are ninety-eight pounds of water in the man of ten stone!

The ocean does almost everything for man. Consider this statement well, and you will be astounded at the way in which we are everywhere dependent, directly or indirectly, upon the sea as the great reservoir of the world's water, and as the manufacturer, by means of its myriads of living contents, of new and useful material from the old and worn-out rubbish, the very refuse and filth, that we daily pour into it. In fact, one

of the principal occupations of civilised man may be said to
consist in making clean water dirty; and one of the greatest
operations of Nature is to make the dirty water clean and pure
again. Like the man in the fairy story, the sea gives us new
lamps for our old battered and bruised ones; and it is mainly
enabled to do this by reason of its immensity and the enormous variety of its population, each able to turn some portion of
our rubbish to account. According to the most recent estimates,
the cubical contents of the ocean is fourteen times greater than
the bulk of the land, and this means that the whole of the land
could be lost in the oceans. Not only so, but if all the conti-
nents and all the islands were dumped down into the Atlantic,
there would still be two-thirds of that great ocean quite clear,
and the whole of the other oceans would be undisturbed. It is
calculated that the entire surface of the globe is 188 millions of
square miles, and of this, the small portion of 51 millions of
square miles represents the land surface, whilst the Pacific
Ocean alone has a surface area of 67 millions of square miles.

It is no wonder that the immensity and mystery of the sea
have always exercised a fascination over man. Emerson de-
clares that "the Scandinavians in our race still hear in every
age the murmurs of their mother, the ocean;" but he need not
thus have limited the thought—in this respect, at least, we are
all Vikings, and the murmurs of our mother still draw us to her
side. Whether we be Scandinavians or Celts, the sea has
power to bring us to her to-day as strong as ever it had over
our forefathers, who found in the seas that lap our little isles
the secret of national liberty, wealth, and power, such as no
other country has ever enjoyed. What a part the sea has
played in the making of the great Anglo-Saxon race! It is but
meet that we should try to understand something of that great
heart of Nature; and for years we have been sending expeditions here and there to sound its depths, and collect facts that
shall one day enable us to know it thoroughly. We cannot
all undertake, or accompany, such expeditions, and must, there-

fore, be content to read with delight of their results; but great numbers of us make our annual pilgrimage to the sea-shore, and, if we will, may learn much of its wonders and beauties without running into danger, experiencing the discomfort of sea-sickness, or risking more than the wetting of a foot.

In the present volume it is the author's desire to act as a friendly go-between, introducing the unscientific sea-side visitor to a large number of the wonderful and interesting creatures of the rocks, the sands, and the shingle beach. Some may think this a work of supererogation, for already many volumes have been issued with a similar object. It is true that there are a number of manuals upon the wonders and the common objects of the shore, but the best are out-of-date or out-of-print, and the recent ones are such shocking examples of bookmaking without much knowledge of the subject in hand, that the practical 'long-shore naturalist smiles and writhes alternately as he turns their pages. Whatever else the present effort may lack, I claim for it this merit, that it has been written in close contact with the things it describes—not only of cabinet specimens, but of the living creatures under natural conditions. There is not a line in the whole volume that has not been written within a few yards of, and in full view of the rocks where the waves forever break, sometimes gently with a low murmuring, almost a whisper; at other times rearing their white crests a mile away, then sweeping across the bay, flinging their malachite curves upon the rocks with giant force and thunderous roar, whilst the foam flakes flying high tap softly at my window.

As far as possible I have dealt with the fauna of the rocky shore separately from that of the sands or the shingly beach, but it must be understood that in Nature the reis a good deal of overlapping. It will also be no surprise to the reader that the rocky shore bulks more largely in these pages than sand or shingle; the rocks with their cracks and caves and pools affording protection to many delicate organisms against the fury of the waves.

Naturalists have marked off the sea-bed into a series of zones, an arrangement which may seem somewhat arbitrary, but which is found very useful in practice. The first or highest of these zones is known as the Littoral zone (Latin, *litoralis*, the shore), and includes all the shore, be it rocks, sand, shingle, or mud, that lies between the highest and the lowest of spring-tide marks. Next to this comes the Laminarian zone, so-called because between very low tide and a depth of about fifteen fathoms of water, the *Laminaria digitata*, or Oar-weed, grows profusely over the rocky ground, and forms a splendid cover for the luxuriant animal life that haunts it. Our district is the Littoral zone, and the Laminarian zone forms our seaward boundary, which we cannot cross, for its exploration needs the use of boat and dredge. It is a very tempting province to enter, for it contains the oyster-banks, and many interesting forms of life.

He who would see the most that the shore has to exhibit to him, must consult the local tide-table, and the table of the moon's changes. If his stay at the sea-side is to be brief, he must endeavour to let the date of his start be governed by lunar considerations. Many business men cannot get away for more than a fortnight, and if any such should wish to make the best use of his time in connection with natural history, we should advise him to begin his holiday at the period of the moon's first or third quarter. He will thus arrive at the time of *neap*-tides; that is, when high-water is low, and low-water not much lower—when, in a word, there is the least difference between high and low water. The local weekly newspaper will in all probability contain the times of high-water for every day in the coming week. If not, he must find out on his first day at what hour low-water is reached, and for at least an hour before that time he must be on the shore with basket of wide-mouthed bottles—glass jam-jars are the best, for they are easily obtained everywhere, and should an accident happen to one through collision with a rock, no great harm is done. Now bear in mind that the time of low- or high-water will be about forty-five minutes later to-morrow than it was to-day, and the same number of minutes must be added on each day to give the correct time for your visit to the shore. Arrived there, it is best to keep close to the ebbing tide, and as it goes further and further back, to turn over the stones and weeds that have just been left by it. In this way you will get acquainted with the best manner of proceeding, according to the peculiarities of the special bit of coast you are on, so that when, a week after your arrival, there comes the spring-tides, you will be able to make far better use of your opportunities than if you had arrived in the locality just at the period of spring-tides.

The lowest tide is the third after New and Full Moon. Then the water goes out to a great distance, and if on a rocky shore you will be able just to step over the border among the Laminaria, and hunt for specimens on its roots and under its long broad fronds. If you really desire to see and find as much as possible with the greatest amount of comfort, then pay attention to your dress before seeking the shore. You should don an old suit of clothes that has become too shabby for ordinary wear. If it is a bicycling outfit, so much the better, for the knickerbockers will be more handy for wading. There is, of course, no necessity for wading, but often it will be found that a "likely" looking rock is cut off from us by a few feet of shallow water, too wide to jump. In such a case wading pays. But it is really best to make up the mind to wade. Take with you an old pair of shoes, and above high-water mark you will find some safe place in the rocks for depositing your walking-shoes, socks, and such other articles of clothing as you wish to doff. Put the old shoes on your naked feet, and roll up your trousers or knickerbockers as high as they will go. You thus run little risk of getting your clothes wet, and your feet will be protected from the sharp edges of newly-fractured rocks and broken shells, or even from the nip of a too familiar crab.

B

Should the idea of old clothes be an objectionable one to you, and you have a preference for something appropriate, I would strongly advise a good knitted Jersey, worn without a coat—at least when the collecting ground is reached. Such a garment is warm without being heavy, and is a protection against the changes of temperature that frequently take place by the sea ; there are no tails to get wet when you sit or kneel on low rocks, and no pockets out of which things can fall when you stoop. For the head a cloth cap is best ; whilst at work wear it with the peak behind, otherwise when you peer closely into a pool it will get wet.

If you visit the village shop or store you can buy for a few pence one of the handy open chip baskets with handle across the middle, that are so much used for gardening purposes. In this you can store your glass-jars, and have them always handy without any lid to open, and can find room for seaweeds, shells, etc. If you are going to the sands you should carry a garden trowel ; if to the rocks, a good strong putty knife with straight edge. You will find in most cases this will do instead of the more cumbersome cold-chisel and hammer that you may *have* to use on special occasions. With it you can separate the upper flake of a slaty rock upon which are desired specimens, by driving the knife in at the edge. For getting anemones off rocks you will find this knife very valuable. In such situations the anemone's base usually rests upon a crust of old acorn shells, sponge, coralline, or other foreign growth on the rock. The edge of the knife should be driven through this crust at a little distance from the desired specimen, and then pushed firmly towards and under it. It will come off with its base— the most delicate part of an anemone—uninjured and undis- turbed, so that when placed in an aquarium it will spontaneously glide off the crumbling rubbish and obtain a firmer footing.

Some of the anemones and other fixed objects in the rock- pools you will find are in too great a depth of water to be got at with ease or comfort ; but by using one of your bottles as a baler you can rapidly reduce the level of the water to a working height. I have in this way almost completely emptied a deep and narrow rock basin, where there was no play for the arms. ·You need have no scruples about destroying a natural aquarium by so doing, for the rising tide will soon put that matter right again. Where I have had to reduce the water in a large pool that would have taken a long time to bale out in this fashion, I have taken down a portable garden pump with splendid results.

In working a "drang" or rock gulley at low water, pay special attention to the lower part of the perpendicular rock-walls, that are most protected from the full force of the waves in stormy weather. Where such a fissure runs parallel with the cliffs, the most productive wall will be that which faces the cliffs, for it is easy to see that in heavy seas this is the part that is protected from the sledge-hammer force of the waves and the big stones with which they batter and bombard the land; therefore, it is the part where soft and delicate organisms have the best chance of flourishing. It will be well also to carefully scrutinise the opposite wall, but when there is only a brief time at disposal devote it to the one we have indicated as the best.

Should you desire to obtain specimens for preservation in the cabinet instead of the aquarium, then you must take a jar of *fresh* water, which should be of a distinctive shape or material, to prevent mistakes Most of the marine creatures are killed by immersion in fresh-water, which has the advantage of not altering their colours, as spirit does in too many cases— notably among the Crustacea. A few of the corked glass tubes that most naturalists use, will be found handy for minute specimens, which are liable to be overlooked if put into the general collecting jars with larger creatures. Overcrowding of the live stock must be avoided, or all will be dead or dying before your collecting is well through.

For small fish, shrimps, and other swimming creatures, you

will require a small net, or rather two nets, for one that is
suitable for catching the small and delicate forms one finds in
the rock pools or swimming near the surface of a smooth sea,
will not be strong enough for drawing through the rough weeds.
The one should be of fine muslin to retain minute forms; the
other should be really a "net," of the very smallest mesh
possible.

On the rocky shore you will find the greatest abundance and
variety of the marine algæ or seaweeds, most of the crustaceans, nearly all the anemones of the littoral zone, a number
of species of fishes, many of the tube worms, the sponges, the
tunicates, and such molluscan forms as the periwinkle family,
the limpets, dog-whelk, tops, slit limpet, smooth limpet, cowry,
and sea-lemon.   On the sandy beaches you will find only such
seaweeds as have been washed in by the waves, shrimps, the
masked crab and the angled crab, launce or sand eel, the
razor-shell, cockle, tellen-shells, horn-shells, the natica, and
other shells.

On the shingle beach little will be found besides empty
shells and heaps of more or less damaged seaweeds, which,
however, are well worth examining, for occasionally one may
find uncommon kinds there, and among them specimens of
animal life.   But it is to the rocky-shore we advise our readers
to give most attention.   The rocks, their pools and crannies,
will engross the attention more, and the harvest will be greater.
By a little local study it will be found that certain winds will
cause the heaping up of certain shells on one particular part of
a beach, whilst other winds bring other things to the same or
different parts.   This knowledge acquired, you will put it to
practical use by finding out what was the direction of that stiff
gale that blew last night, and then bending your steps in a
particular direction, you will be able to take your pick of the
shells before the hinges have become broken and the valves
separated.   There are many species of mollusca whose shells
you will only acquire in this way, unless you are able to go

dredging, and thus get up the living creature from the sea-bottom. All such shells, though they may look perfectly clean, should be carefully washed in *fresh* water, to get rid of the salt, that would otherwise hang about them, and prevent them becoming absolutely dry, as cabinet specimens should be.

Probably, after you have really seen something of the exceeding beauty of the rock pools and the little marine caverns, you will be fired with the ambition to start a small marine aquarium when you return to your own home. You really ought to be filled with this desire a month or so before you seek the shore, so that you might provide a suitable vessel or vessels, and allow the sea-water to settle down and the contained germs of vegetation to start into active life, and so be ready to support animal life. We will suppose you have made some provision of this sort before leaving home, and now desire a suitable selection of creatures to fill it. My advice is, be modest in scheming, and for a first experiment start with creatures that consume very little oxygen—you cannot have better subjects than the anemones. These should be conveyed not in water, but each specimen wrapped lightly and separately in soft weed, and the whole packed in more weed in a light wooden box. The pools should be searched for a rough, uneven piece of rock, upon which small *green* weeds are growing, and this should also be placed in your aquarium as a suitable base for the anemones. Most marine animals travel better in weed than in water, which rapidly becomes foul in travelling, and destroys all that have been entrusted to it.

CHAPTER II.

LOW LIFE.

SOME persons go to the seaside every year for several weeks, and yet know little of its treasures. Take away the bands, the bathing machines, the itinerant entertainers of various kinds, the bustling crowds that pass and repass on the grand parade, and they are lonely, miserable, with nothing to occupy their minds. Of the illimitable sea, the cliffs, the sands, the passing sails they soon tire. For a very small sum, as money is considered to-day, such a person could acquire a tolerable microscope, and a very little application to books would put him in the way of getting an absolutely endless fund of interest, knowledge, even amusement from it. Through the magic glasses he enters another world; or, rather they enable him to see that other half of Creation with which he has been rubbing shoulders all his life, yet without seeing the creatures. With such an instrument and the knowledge how to use it, a man may defy the demon *ennui* wherever he may be. With such an instrument at home a person who is not a naturalist may be induced to look into a rock-pool, to take samples of its fauna and flora, and by and by to become a naturalist without intending or knowing it.

Behold how easy a thing it is! He has but to take away a phial full of the water, a tiny bunch of coralline, the finer green weed, or a snippet of sponge from the walls of the pool, and he has abundance of material whose marvellous beauties of form and colour will delight and astonish him when he has had time to examine it under the microscope. For the coral-line tuft and the lowly weed, when washed out in the sea-water, will yield him multitudes of Infusoria, Rhizopoda, and the infantile stages of many of the higher groups of life.

The Foraminifera are the minute creatures which have so largely contributed to the formation of the enormous beds of chalk we find in Surrey, Kent, Sussex, and other counties, such as the explorations of the *Challenger* showed us are being formed in the deep sea at the present time. So minute are they that one hundred and fifty of them placed side by side would not measure more than one inch, and of such insignificant creatures the chalk is almost entirely composed. What are they? How are they fashioned? How do they live? These questions probably occur to the reader, and I must do my best to briefly answer them.

There is a minute creature, plentiful in ditches and similar accumulations of stagnant water into which decaying vegetation has fallen. It is a minute speck of animated jelly, without form, substance, or limbs. There is, in fact, no closer analogy than the speck of almost clear jelly, to which in some mysterious way life has been given. In the words of the late Dr. W. B. Carpenter, who made a special study of these creatures: " A little particle of homogeneous jelly* arranging itself into a greater variety of forms than the fabled Proteus, laying hold of its food without members, swallowing it without a mouth, digesting it without a stomach, appropriating its nutritious material without absorbent vessels or a circulating system, moving from place to place without muscles, feeling without nerves, propagating itself without genital apparatus, and not only this, but in many instances forming shelly coverings for symmetry and complexity not surpassed by those of any testaceous animal."

With the exception of the last three-and-twenty words the above description refers to the *Amœba* and its allies; but in the Foraminifera we have a sort of advanced type of amæbæ,

* It is now known that this jelly-like material is not of so simple a character as was supposed a few years since ; the most modern microscopes prove it to be not devoid of structure.

a more æsthetic race that have taken to build themselves
houses, in most cases of graceful form, such as are referred to
in Dr. Carpenter's concluding words.   One of the fresh-water
amæbæ is named *Difflugia*, and it distinguishes itself by coating the greater part of its small body with particles of sand
and other matter picked up as the *Difflugia* rolls along.   The
Foraminifera do not resort to so clumsy a method of satisfying
their architectural instincts.   In the course of their feeding
they take into their primitive systems a good deal of carbonate
of lime, and instead of casting this out as innutritious, useless
stuff, they secrete it as shell, in many cases not unlike the
shells of mollusks, but with minute pores (*foramina*) all over
them.   From this character they derived their name *Foraminifera* or pore-bearers.

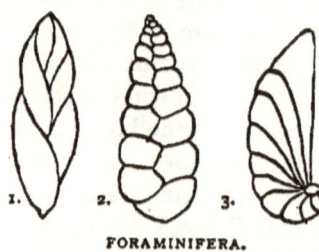

FORAMINIFERA.
1. Polymorphina. 2. Textularia.
3. Cristellaria.

Within these perforated shells
live the amæba-like animals, and
through all these minute pores
they protrude still more minute
threads or wisps of their living
jelly to use as limbs wherewith to
pull themselves along, and to
catch their food.   There is a very
ancient conundrum which asks:
"What is smaller than a mite's
mouth?" a mite being formerly
considered to be the least of all animals and a very
minute thing indeed; therefore, to imagine the mouth of a
mite was to conceive of something so very small as to be
almost beyond conception.   But then came the answer:
"That which goes into it!"   Of course, if a mite had a mouth
it must have it for the purpose of eating, so that though
nothing were known smaller than a mite, yet a mite must have
a mouth, and that could scarcely be quite as large as the mite,
and its food must be smaller than its mouth.   A naturalist
would say that this line of reasoning is weak, and it undoubtedly is so, for there are creatures that contrive to swallow

things that are much larger than their mouths; but there is
no occasion to split hairs just now. These Foraminifera are
in some cases invisible to the unassisted vision, but as each
is pierced with many pores, it follows that the individual pore
must be almost inconceivably small, though still smaller are
the wisps of jelly that protrude through them and invest the
outside of the shell. For it must not be supposed that these
structures are secreted like the shell of the snail, that the
animal may live within it; rather it is like our own skeleton,
built up within our bodies.

Some of these shells have but one chamber, like *Lagena*,
which is flask-shaped, and *Entosolenia*, in which the long neck
of the flask has been pushed down inside the globose portion.
Others have many compartments, but these are subject to
great variety of arrangement, each species having its own
special form. *Dentalina* has the chambers placed one behind
the other in a straight or curved line. In *Nonionina*, *Polysto-
mella*, *Rotalina*, *Globigerina*, and others they are rolled in a
spiral, and resemble the chambered shell of the Nautilus; or
they may be twined, *not* spirally, round an axis, each making
a half-turn.

In some respects similar to the Foraminifera are the Polycis-
tina, which are equally minute creatures, whose skeletons are

POLYCISTIN.

of flint instead of chalk, and the perforations
are so large and so close together that the
term pore no longer adequately expresses
their proportionate size. They are more like
windows, but with little intervening stone-
work. The jelly-substance, called *sarcode*,
flows out through all these windows in the
form of threads (called *pseudopodia* or false
feet) as in the Foraminifera, spreading over
the outer surface and acting as legs and arms
by means of which the creature moves and
captures its food. They feed upon infusoria
of various kinds, and the diatoms and desmids,

which appear to be paralysed by contact with the pseudopodia. They also seem to derive part of their nutriment from the exertions of some minute yellow-bodies, a species of algæ (*Xanthellæ*) that are lodgers within their substance. These lowly plants, which have sometimes been incorrectly alluded to as parasites, elaborate starchy products by the aid of their chlorophyll, and on their death this material is available for the nutrition of the Polycistin, which also can make use of the oxygen given off by the plant.

There is one of these low forms of life in which almost all visitors to the sea-shore take an interest—or rather they are interested in certain signs of its vital activity—the mysterious phosphorescence of the sea. There is no moon visible, the sea is quiet, and our reader late in the evening takes a stroll along the edge of the waves, "before turning in." He is charmed to see the ripples as they break upon the shore brightly outlined with glow-worm light, and stays long to enjoy the elfish illumination. Now my advice is, do not stay long, but hasten back to your "diggings" and get a bottle; then return and fill it with sea-water at a spot where the phosphorescence is most abundant. You can then examine the creature that produces the strange light.

If now you continue your stroll along to that part of the sea-wall where the male villagers most chiefly congregate to spin yarns with a more or less saline flavour, and to discuss village politics, you will probably hear them talking about fishing prospects, and if it is in early summer, mackerel will be in their talk. "Well," says one, "there's no doubt the fish are about, and I propose that we get the sean-boats ready, and to-morrow night we'll try the briming." The meaning of which dark saying is that to-morrow evening they will row across the bay till they come under the shadow of the great headland, and there they will adapt the focus of their eyes to seeing below the surface of the crystal waters, and watching for the streaks of phosphorescent light that break from the fins and tails and scales of the mackerel as they pass through the sea.

This light is the "briming" of the fishermen. It is due to the movement of the fish exciting the light-producer, just as in a marine aquarium in a dark room you can produce a similar effect by blowing the surface of the water into ripples. The six long oars of the big sean-boat every time they dip into the water send a spray of light into the air, and as they again leave it a shower of glowing pearls drops from each. The prow of the boat sends up a fountain of pale heatless fire on either side, and an ever-widening track of the same mysterious light marks the way the boat has come.

All these brilliant effects are produced by millions of a tiny Infusorian, individually so small that twenty of the finest specimens, placed closely together in Indian file, would only produce a procession one inch in length, whilst of mediocre examples it would require from fifty to eighty to cover the same space. Its size may be insignificant, but it has a name which will at least inspire respect with some persons—*Noctiluca miliaris*—which may be Englished as the Sea Nightlight. If now we go together to your lodgings and examine that bottle of sea-water with a lens we shall be able to make out a large number of these creatures swimming about, and by means of a pipette or dipping tube we can isolate a specimen and place it under the microscope. There it is revealed to us as a peach-shaped individual, the spherical mass being partly mapped into two lobes by the slight groove that, as in the peach, runs down from the depression in which the stalk is attached. The stalk in this microscopic night-light is represented by a long flexible tentacle, or *flagellum*, by means of which *Noctiluca* moves through the waters, much as a fisherman will propel his boat by the skilful use of a single paddle at the stern. There is a shell-like envelope of transparent material through which may be seen a meshwork of granular material, denser than the body-mass. A funnel, opening near the *flagellum*, becomes lost in this granular matter; this is the creature's mouth and gullet, within which lies a smaller flagellum. The gullet simply opens into the central proto-plasm; no continuing alimentary canal has yet been made out. Reproduction is effected by several methods: one is the division of the creature transversely into two, each complete, but for the time smaller; a second method is the conjugation of two individuals and the subsequent breaking up of the pro-toplasm into numerous spores, each provided with a flagellum. But this breaking up process may occur independently of conjugation. The spores move by the lashing of the flagel-lum, and gradually develop into the adult form. The light is produced in flashes just under the clear cell wall, and pure sea-water, rich in oxygen, is necessary for its continued brilliancy. At times, on summer evenings, *Noctiluca* is ex-tremely abundant in the littoral zone, and it is then impos-sible to take up a glass of water without getting thousands of specimens.

If you occasionally indulge in boating, many forms of low-life, or the larval condition of higher forms may be obtained without difficulty. Take a piece of thin, round cane—about the thickness used in training a child in the way he should go —and bend it into a hoop. The two ends should be cut half through for an inch of their length, so that their flat surfaces can be brought together and secured by several turns of a piece of thin copper wire. Now to this cane secure a small flat piece of lead, so that when thrown into the water the hoop will assume an erect position. If you should have a couple of inches of "compo" gas-tubing handy, this will do admirably, and may be slipped over the cane before the ends are lashed together. Upon the hoop now stitch a muslin bag to serve as a net; and to three or four equi-distant points on the frame attach short, strong strings of equal length, and join their ends to a length (say three fathoms) of fishing line. This may be made fast to one of the thwarts of the boat, or held in the hand, whilst the net is thrown overboard. The movement of the boat will cause the net to collect a large number of minute creatures that float on the surface or immediately below it. From time to time it should be hauled in, and the bag turned inside out and washed in a glass jar of sea-water. In this way many interesting forms may be secured.. A calm, sunny afternoon should be selected for this work, and the boat should be rowed gently.

CHAPTER III.

SPONGES.

To many persons the statement that we are going for a ramble among the rocks in quest of sponges will merely suggest the idea of wreckage, and they will suppose that we have had information that a vessel, part ot whose cargo was Turkey sponges, has gone to grief on the rocks near, and that sponges are to be had for the trouble of picking them up. And should they venture to accompany you on so promising an expedition, they would certainly consider you demented as, having reached the rocks that are only uncovered at very low-tides, you proceeded to point out the green and orange and brown and whity-yellow expanses that coat the vertical faces of the rocks.   All these things to them bear no resemblance to the only sponges they know—the ones they use daily for purposes of ablution. You can show them something approaching nearer to their ideal, if you hunt among the thick stems of the shrubby weeds on the rock.   There, encrusting a branch, is a yellowish-brown form with rough surface and large pores very much like those they know all about. And attached to various weeds are others of the shape, size, and colour of melon seeds, with porous surface and open end.

Your friend, though disappointed, maybe, that he is not to share in the salvage of some splendid bath sponges from the supposed wreck, cannot help feeling some interest in the extensive layers of colour on the rocks, some of it raised into conical hillocks, and suggesting a fairy plain thickly studded with volcanoes. You tell him that these are really aquatic volcanoes so to speak, and that if you could get a portion off

Halichondria
panicea.

Grantia
coriacea.

Halichondria
incrustans.

Leuconia
nives.

Hymeniaci-
don
albescens.

Pachyma-
tisma
johnstoni.

Leuconia gossei.

SPONGES.

the rock, you could exhibit the phenomenon to him at work in a shallow dish of sea-water. Thereupon he, thinking to be of service to you tears off a slice of the pale green, hillocky sponge *(Halichondria panicea)* and breaks it up hopelessly. However, we will turn his clumsiness to account and take a view of the interior thus violently exposed. We see that these crater-like openings are the outlets to tubular spaces running through the sponge, and from these passages smaller branches go off at right angles, whilst these and the larger openings are surrounded by tissues that are very like bread in consistence ; and that is really only a way of explaining that they are spongy. Now the whole of the substance of these sponges, as you may see by microscopical examination, is composed of myriads of minute flint spicules, finer than the most delicate fragments of " spun-glass," and of beautiful forms. Some are simple rods, straight and curved ; others forked at one end ; some like a gribble ; others what is known as quadriradiate in form.

. Now in some species these spicules are not arranged in any order ; they are merely jumbled together, and their remarkable forms make it easy for them to become entangled. When so en-tangled they form the skeleton of the sponge. Each sponge is a co-operative colony containing many thousands of members, and these are represented to our view, through my pocket lens, in the mass only, as a thin clear jelly investing the spicule-tangles, or rather the spicules are imbedded in the *sarcode* as this living matter is termed. If we were to chip off a thin flake of rock with its investing sponge intact, and place the whole in a glass vessel full of water, we could observe the movements which manifest its vitality. A little finely powdered indigo or other colouring matter should be dropped into the water near the specimen. On closely observing it would be seen that many of these minute granules were flowing towards the sponge, then that they entirely disappeared through the very fine openings in the surface. A little later these particles

c

will re-appear, not where they went in, but in a denser stream issuing from one of the craters, which are scientifically designated *oscula* to distinguish them from the minute pores.

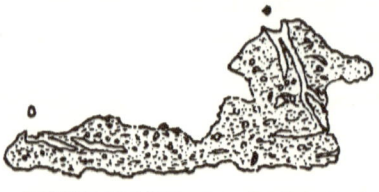

SECTION THROUGH CRUMB-OF-BREAD SPONGE.

If we dissect the sponge under a microscope, we shall find that from one of these *oscula* a broad passage runs through the centre of the mass, and from the walls of this the minute pores run off to the outer surface. This central cavity is invested by a living membrane which, when examined through a higher power of the microscope, is seen to consist of myriads of organisms closely packed together side by side, and each resembling a glass vase, spherical below, with a wide neck, and from its centre there issues a long antenna-like process. This is called the *flagellum* (Latin, a whip), because its office is to lash the water. These flask-like organs, with their flagella, present a wonderful likeness to some free infusoria known as collared monads, and over this likeness and all that it may or may not imply to the systematic naturalist much ink has been shed, and the sounds of contro-versial strife it engendered, though now faint, are still audible. Into *that* question we do not go.

The combined lashing of these little whips in unison sets a strong current of water flowing through the central passages and out at the *oscula*. To feed this stream, water flows in automatically through all the little pores, and brings with it the infusoria and other minute particles of life with which the sea is swarming. These come in contact with the lips of the flasks in the interior over which the living jelly of the sponge is steadily flowing. The infusoria flow with it and are carried away by the current to a little clear space *(vacuole)* in the lower part of the flask, where it is digested, and the refuse portions are thrust out to go in the general stream and be carried out through the oscula. Each of these cells may be taken therefore as a separate individual, enjoying home rule, yet taking part in general efforts for the whole sponge-community, for we find that by some strangely communicated understanding, all these cells cease lashing the water for a time as though resting (or digesting their food), and the craters cease to pour forth their streams. But then after a time activity is resumed, the craters belch forth again, and we know thereby that the flagella are in active operation down below, not merely capturing and digesting food, but also absorbing oxygen from the inflowing streams, whereby vital energy is maintained.

After the cells have become full grown, they split transversely or longitudinally, and so increase their number, which means that the size of the colony increases. But some of these divided portions develop into eggs, which after fertilization are swept out into the ocean by the outflowing current, and settling upon some rock become glued down and grow, gradually, by division and subdivision, producing a new colony. Such is a highly condensed account of the general phenomena of sponge life. There are variations upon it in the life-history of well-nigh every species; but this will suffice to give my reader a general idea of what sponges are. For the rest, he must go down among the rocks, and search out the various species of many forms, and endeavour to add to the general sums of knowledge by some fresh observations respecting British Sponges.

However startling the statement may sound, there is no lack either of specimens or species on the British coasts. Some of the most conventionally sponge-like of these must be sought by the dredge in deep waters, but our own hunting ground, the rocks that mark the shoreward-bounds of the laminarian zone, if carefully inspected at low spring tides, will afford more specimens in half-an-hour than we can exhaust the interest of in a week. That this is no mere figure of speech you will agree when I add that Dr. Bowerbank published a work in three volumes dealing only with British Sponges, and to these a supplementary posthumous volume, edited by Dr. Norman, has since been added.

Where the rocks rise high above the shore with their upper portions tilted towards the cliffs, we shall find several species incrusting the vertical or overhanging surfaces of these rocks, such as *Halichondria incrustans*, whose buff-coloured bread-like surface is diversified with slightly raised oscula. Its principal spicules are knobbed at one end, in which respect it differs from the similar *Halichondria panicea* which is peculiar in having only one type of spicules—a rounded rod, slightly curved or quite straight, but pointed at each end. Ellis called this species the Crumb-of-bread sponge, a name which is reflected in the scientific cognomen *panicea*. It is one of the most plentiful of the encrusting species, and may be readily known by the greenish-yellow or distinctly green colour of its extensive patches.

Not far from the Crumb-of-bread will in all probability be found the similar Sanguine sponge (*Halichondria sanguinea*), of a bright red colour. The conical elevations of the *oscula* in these species distinguish them readily from the plump, though narrow bands of *Microciona carnosa*, a plentiful species that creeps extensively between the other kinds, its pale red branches being very unequal in width, and alternately contracting and swelling out, joining and separating. This will be found figured in the lower left-hand corner of our illustration on page 29.

A very noticeable species on account of its neat compact shape will be found attached to various red seaweeds, with which its whitish colour contrasts well. It is a small oval, usually from a quarter to an inch in length, very flat, but yet hollow, with a large vent at the free and larger end. This is the *Grantia compressa*. Careful search among the indescribable medley of "unconsidered trifles" that crust the rocks beneath the shelter of the Fucus-growth, will reward us with a little spherical sponge with tubular oscula at the summit formed of spicules, and its general surface bristling with long spicules. This is the *Grantia ciliata*, looking like a little gooseberry.

GRANTIA COMPRESSA.   GRANTIA CILIATA.

There are many other forms, for which I must refer my readers to Dr. Bowerbank's work, where also will be found descriptions and figures of many deep-water species, such as the more conventional sponge-like *Chalina oculata*, in branching masses nine or ten inches high.

There is, however, one other we must mention; the so-called Boring-sponge *(Cliona celata)*, which attacks various shells and stones. It is quite a common occurrence for the rambler along the shore to pick up the shell of some mollusk, and find it so tunnelled, the borings branching in every direction, that what would otherwise be as strong as stone is now as weak as poor strawboard, and will yield to very slight pressure or strain. On breaking such a shell across we get both cross and longitudinal sections of these tunnels and chambers, and find some of them to be lined with a dark-brown filmy tissue, the remains of some past inhabitant; others contain portions of this *Cliona* sponge, living or dead; others again contain little bivalve shells that just fit the aperture, whilst yet another set exhibit clean walls that may not have had any animal inmate. Much controversy has raged over the question whether these excavations have been made by the sponge, or by some boring worm, and there have not been wanting as advocates of either view men whose authority on sponge matters is unquestioned. Where such doctors differ how shall humble observers venture to give a verdict? For my part, I cannot give my support to the contention that the sponge has bored the clean holes, hollows, and tubes that I have seen in the large numbers of attacked shells I have broken; neither am I prepared with an opinion as to the creature that did make them. I believe that on this matter, as on many others connected with natural history, we have much still to learn, and every student of Nature should have his eyes and his mind ever open to receive hints from Nature herself as to her methods. One of these days, some lonely wanderer by the margin of the wave will show us how simply this boring is accomplished, and we shall all wonder that we never thought of the possibility before. But whatever views or lack of views we may have upon the question, "who made the burrows?" there is no doubt that the sponge does exist in some of them, and its spicules embedded in the yellow sarcode are well worthy of minute observation.

CHAPTER IV.

ZOOPHYTES.

NOT many years ago our knowledge of the lower forms of life was very imperfect, and it was believed that the gulf between the animal and vegetable kingdoms was bridged over by certain creatures which could not properly be classed in either, because they appeared to unite the characters and organization of each. Such was the case with the sponges, already dealt with, and with the creatures now to be considered. These last were on that account called Zoophytes, or animal-plants, a term which we must render to-day as plant-like-animals. Some of us have again got to the notion that there is no sharp division between animal and plant-life; but with increased knowledge we have put back the debatable or common ground much lower in the scale of life.

With the whole of the families included in this division of life, I do not propose to deal in the present chapter: the Sea Anemones and the Sea Jellies, for instance, being treated in succeeding chapters, for each group deserves and demands a chapter to itself. It is characteristic of the Zoophytes that they form a bag of jelly-like material, with an opening at one end which may be regarded as a mouth, though it is without tongue or teeth, and opens directly into the stomach. Around this mouth are set a number of limb-like organs, called tentacles, which are used for seizing the prey and conveying it within the orifice. Their entire structure is very simple, and apart from primitive muscular and nervous systems, and the possession of stinging threads, which can be quickly extruded through the exterior walls of the body, they appear to be almost innocent of organs. This form of structure is generally referred to as a Polypite, and its appearance has been made familiar by the descriptions and figures of the Hydra or Polyp of our stagnant fresh-water ponds. From their general agreement in structure with the Hydra, the creatures to which much of this chapter will be devoted, are called Hydroid Zoophytes. There are, however, but few species that occur solitarily, like the Hydra. In most cases they are associated in inseparable colonies. The egg of a zoophyte gives rise, it is true, to an organism resembling Hydra, but this individual does not long remain solitary; it produces many buds, which rapidly develop, and in turn produce other buds, so that before long there is a colony that may number its thousands of polypites. However numerous the individuals may be, we may be sure that the colony has been the production of a single egg. One came from that egg, but all the others were produced vegetatively by budding from the original polypite, or as later generations from such bud-originated polyps.

A slight examination of such a colony will show that the polypites themselves are held in association by an investing substance (*cænosarc*), which takes the form of a living tube of thin flesh, which adheres to rock or shell or seaweed, acting as a support for the community, and also reproducing the polypites. It consists of two distinct layers, an inner and an outer, and sometimes there is a third layer of a different kind between these two, muscular in character. In most cases the outer wall of this tube secretes a sheath of a substance called *chitin*, of which the external skeletons of insects are composed. This sheath is known as the polypary, because into it the individual polypite withdraws itself. It is this polypary that the sea-side visitor finds attached to weeds or shells, and concludes, from its moss-like aspect, it is a seaweed, and probably adds it to his collection as such.

Now if we get down among the rocks near low-water, and look among the coarse brown weeds, we shall not look long before we find one whose stem and parts of the frond are covered with a plantation of erect-growing "somethings," that look like the backbones of some small fishes. They are only about an inch in height, very slender, and regularly notched on each side. Some of the specimens have one or two branches, but most of them are simple erect stems. It is known as the Sea Oak Coralline (*Sertularia pumila*), and if we examine it with our lens, we shall find that each of the notches represents the space between the elegant crystal vases that are arranged symmetrically along each side of the stem. These vases are known as *calycles*, and in each there stands a polypite, reaching out its upper portion and waving its tentacles. In case of danger the polypite can be withdrawn into the calycle; and certain species have an automatic contrivance for closing the mouth of the vase when they have retreated within. All genera have not these calycles.

SEA OAK CORALLINE.

Returning to the animal for a moment, it should be explained that its organization is so low that there is no true circulatory system for the renewal of the body, by the carrying of elaborated food from the stomach to distant parts of the body; but by the activity of innumerable eye-lash-like hairs on the surface the whole of the particles of food digested in the stomach are carried all over the system to be then assimilated by different parts.

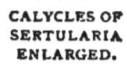

CALYCLES OF SERTULARIA ENLARGED.

Within the circle of tentacles is the mouth, which is sometimes cut into lobes, and is generally borne upon a very mobile proboscis, which may be withdrawn or protruded, and in some genera takes a trumpet-shape; in others it is conical. In the winter the cœnosarc may frequently be found with all its calycles empty; and it might then be supposed that the zoophyte is dead and only its skeleton remains. But this is not necessarily so, and a closer inspection may convince us that the organism is alive. In spring it will furnish its calycles with new polypites, and all will go merrily again. At certain seasons buds of a peculiar structure are formed, which develop into polypites, whose function it is to produce eggs, instead of catching and digesting food for the colony. These are known as *gonophoræ*, and sometimes they remain where they were produced, simply bursting to discharge their contents. In other cases they detach themselves from the parent colony at a certain stage in their development, and float off, having all the appearance of minute jelly-fishes. Some of these, instead of remaining small, attain an enormous size, so that it is difficult to credit their origin to the so-called coralline upon which they were produced, and of which in turn they are really the egg-bearers. The eggs they scatter will develop into plant-like growths such as they were produced by; from the edge of their jelly-umbrella and from its handle, buds are given off, which open as jelly-fish like itself.

Growth proceeds rapidly among these creatures, and if a balk of timber be immersed in the sea, it is not long ere there is a fine forest in miniature upon its surface, and that forest will consist of some of these corallines. The species are generally distributed along our coasts, but a few are local. Thus the finest of all the British species of *Sertularia—Diphasia pinnata—*is found only on the coasts of Devon and Cornwall, where most other species attain their maxima of beauty and luxuriance. Its relative, *Diphasia alata*, as well as *Calycella fastigiata* and *Aglaophenia tubulifera*, have been found in Britain, only in Cornwall, Shetland, the Hebrides, and on the west coast of Scotland. On the other hand, certain species belong to the north, and such species as *Salacia abietina*, the Sea-fir, and *Sertularia tricuspidata* are not found on our shores below the north-east coast. *Sertularia fusca* is similarly confined, so far as our seas are concerned, to the east coast of Scotland and the north-east of England; and *Thuiaria thuja*, found on the east coast, is rare in Devon and Cornwall; whilst the species of *Aglaophenia* are plentiful on our south-west and north-west coasts, and rarely seen on the north-east.

Although some species are distinctly deep-water forms, necessitating the dredge for their capture, the vast majority inhabit the littoral and laminarian zones. Among the littoral species are many of the rarer forms, and some of these are found only on special species of seaweeds, or on the shells of particular mollusks. Mr. Hincks, whose beautiful work on the "Hydroid Zoophytes" you must see, gives some very good advice as to collecting in the littoral zone. He recommends my own favourite plan of lying flat beside the rock-pool, and bringing the eye close to the water. "He should bring his eye to the edge of the pool, and look *down* the side, so as to catch the outline of any zoophytes that may be attached to it amidst the tufts of minute *Algæ*. He must not be content with a hasty glance, but look and look again until his eye is familiar with the scene, and may accurately discriminate its various elements. And let him watch for the *shadows;* for in following them he will often secure the reality. I have frequently detected the tiny *Campanulariæ* and *Plumulariæ* in this way, by means of the images of their frail forms which the light had sketched on the rock beneath them. For tools, the hunter must have his stout, flat, sharp-edged, collecting knife, a long-armed and substantial forceps, and a varied array of bottles, ranging from the homœopathic tube to the pickle-jar. If his choice of ground be good, and his patience proof, and his eye quick, he will have an ample reward for his labour in the rich spoil of beauty which he will bear away, even if he should not hit upon any novelty; but amongst the minute zoophytes there is still, I have no doubt, much to be done in the discovery of new forms, as there certainly is in working out thoroughly the history of those that are known."

I hope that in the foregoing remarks I have made it quite clear that our Sea-oak Coralline is not an individual but a community of individuals—a community on the strictest of co-operative principles, in which the good fortune accruing to one of the polypites by food falling in its way, is shared by all alike; for a polypite cannot digest it and retain it to its own selfish use, instead, it goes to the nutriment of the common-wealth.

Some of these Hydroid Zoophytes, though sharing the communist character, are much simpler in form, and we shall find a common example ready to our hand on almost anything in the way of stone or shell removed from a rock-pool. It is a minute creature, as stout as a "short white" pin, and about a third of the length, white or pinkish; a number of them spring in a row from a creeping stem of firmer substance, in which are well-defined tubular openings, in which the upright bodies stand. These answer to the calycles of *Sertularia*, just as the upright bodies agree with the polypites of that genus. The name of this creature is *Clava multicornis*, and it may conveniently be called the Many-horned Club. It gets its name *Clava* from the shape of the polypite which thickens towards the top, and then tapers off again to the summit, where its mouth is situated. It has a number of tentacles, varying from ten to forty, according to age, but these do not form a regularly-disposed crown round the mouth; instead, they are placed anyhow on the thickened part of the polypite. The name *multicornis* refers to these many-horns or tentacles. An advance on this type is seen in *Coryne pusilla*, a much larger but equally common inhabitant of our rock-pools, in which the tentacles are knobbed, and are arranged in a series of more definite whorls.

There is another group which is more likely to be confounded with the Sertularians by those who are content with hasty glances at things; but species of the one group may be readily distinguished from the other by the aid of a simple lens.

The Sertularians, as we have seen, have the calycles arranged symmetrically on each side of the axis. The Plumularians, as the other group are called, have their calycles arranged along one side only of stem and branches. The Sertularians are frequently spoken of as Sea-firs, the arrangement of the calycles giving some species a very close resemblance to the branches of fir-trees. In the Plumularians, the resemblance much more nearly approaches a feather.

PLUMULARIA PINNATA.

Hincks, describing *Plumularia cornucopiæ* says:—" In the present species a conspicuous band of opaque - white encircles the body, like a girdle, a little below the tentacles, and adds much to the beauty of a colony in full life and activity, when its many polypites are in eager pursuit of prey, stretching themselves forward, and casting forth their flower-like wreaths, now suddenly clasping their arms together, and then as suddenly flinging them back ; now holding their motionless, the tips elegantly recurved, and then on some alarm shrinking into half their size, and folding them together like flowers closing their petals when the sun has gone."

PLUMULARIAN,
PORTION ENLARGED.

In addition to the calycles in which the polypites live, there are special reproductive chambers as in the Sertularians. In this species (*P. cornucopiæ*) " they assume the shape of an inverted horn, and are formed of material translucent as the finest glass. Each one of them, in fact, is a little crystal cornucopia, in which is lodged one of the reproductive members of the commonwealth, a class totally distinct from that which is charged with the function of alimentation. These graceful receptacles are several times larger than the calycles, from the base of which they spring, singly or in pairs, and within them the ova are produced and the embryos matured which are to give rise to new colonies."

One of this group, the Lobster-horn or Sea-beard (*Antennularia antennina*), shown at the back of the illustration of acorn shells on page 183, has the calycles arranged in whorls all around the axis, which produces a very singular appearance, not at all unlike the antennæ of some of the larger crustacea.

In the Creeping Bell (*Calycella syringa*) so common on seaweeds, etc., the calycles are more bell-shaped, and the mouth of the bell is fringed with a series of large triangular teeth, similar to the *peristome* of many moss-fruits. When the polypite withdraws into his calycle, these teeth bend inwards, and so close the opening.

Many of the forms of Jelly-fish to be described in the next chapter, though they are described with separate names, are now known to be merely stages in the history of some of these Hydrozoa or Hydroid Zoophytes—the developed free-swimming gonophoræ previously mentioned.

A singular member of the group has the form of a jelly-fish, but does not act as one. This was formerly named Lucernaria, but is now known as *Haliclystus octoradiatus*. It was thought to swim like a jelly-fish, but it really creeps. Its form is like a ladies' sunshade that, instead of being the ordinary umbrella shape, tapers off to the stick at the top. What would be the ferrule of the sunshade is the footstalk of *Haliclystus*. By this footstalk it attaches itself to a weed, say, and hangs down its eight arms with their connecting web, and by means of a little knob on the edge of the web alternating with its "arms," it is able to take hold until it has "looped" like a geometer caterpillar, by bringing its footstalk forward and taking fresh hold. The extremities of the eight arms (or ribs

HALICLYSTUS.

of the sunshade) are ornamented with tassels of tentacles, and it uses these after the manner of a sea anemone when it wishes to secure food. It, in fact, has some of the peculiarities of both jelly-fish and anemone, though it will not act quite consistently with either character. I have found it on Laminaria and other weeds at low water, and a few months since I picked one off the plumage of a dead guillemot, that had been drowned in a storm and afterwards washed ashore.

There is an important group of incrusting organisms that you will find represented on almost the first specimen of *Fucus* you pick up, and which you may be tempted to class with these zoophytes; but they occupy a much higher position in the scale of life. I refer to the Sea-mats, the Sea-scurfs, the Bird's-head Coralline, and allied forms, whose proper designation is Marine Polyzoa. They are more nearly allied to the mollusks, the structure approaching towards that of the Lamp-shells. They are associated in colonies (*zoaria*), but there is no connecting *cœnosarc* as in the Hydrozoa, although there is communication between the chambers by wisps of animal matter. Each chamber of the Sea-mat marks the habitation of a complete individual, who catches, eats, and digests for himself alone, not for the colony. These chambers are of a horny, persistent character, secreted of course by the poly-pide; with a small opening through which the creature protrudes its mouth and fringe of tentacles. Its body consists of a thin bag filled with a clear fluid, in which can be traced the gullet enlarging into a simple stomach, contracting again into the intestine. There are muscles by means of which the upper part of the sac with the mouth and tentacles are with-drawn inside the lower part Add to this a nerve-ganglion

beside the gullet, sexual organs within the sac, and the polypide is fully described.

The original founder of the colony was produced from an
egg, and was for a time a restless larva, swimming and creeping and whirling around by the aid of *cilia*. Finally it settles
down on weed or stone, and becomes anchored; drops its
cilia and develops its horny chamber and its crown of tentacles. Having reached its full degree of growth, it buds at
the sides, and originates other creatures like itself. Just as
the solitary daisy root or chrysanthemum throws out what the
gardener terms suckers, and soon becomes the centre of a
clump of similar plants; so the solitary Sea-mat soon becomes
only one in a symmetrically arranged colony, containing
hundreds of individuals, all produced by budding from the
original egg-produced polypide.

Some of these colonies have a number of queer adjuncts,
which bear a startling likeness to the head of a bird of prey,
with moveable jaws, that are for ever snapping. These have,
of course, given rise to many theories to account for them;
but it appears now to be generally accepted that the " bird's
head " is a specialised member of the zoarium who serves
some purpose, prob-
ably of defence, or of
scavenging, that is of
advantage to the whole
colony. In some spec-
ies, this differentiation
of individuals takes
the form of a long
whip-like process, con-
stantly lashing, instead
of the snapping jaws.
The forms of the marine
polyzoa are very varied,
but we shall be unable

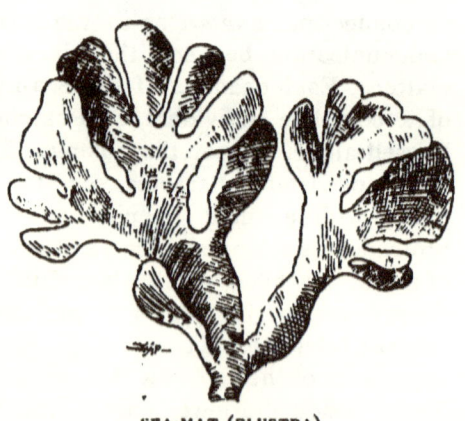

SEA-MAT (FLUSTRA).

to do more than indicate a few of them here, leaving the reader to make wider acquaintance with a most interesting group by studying the species in Hinck's *British Marine Polyzoa*.

The Sea-mat (*Flustra foliacea*) is a deep-water form, whose colonies take the shape of fronds, resembling *Fucus serratus* in outline; but it is thrown up on the beach in great quantities, and it will be one of the first things you will find on the shore, especially if you rout about among the weeds washed up by every tide. Creeping over these flat frond-like masses you will probably find other species that take a more branching form, such as the common Creeping Coralline (*Scrupocellaria reptans*), or the more bushy Bird's-head Coralline (*Bugula avicularia*). The Tufted Ivory Coralline (*Crisea eburnea*) has tubular chambers of ivory whiteness; it is of branching habit, and occurs on some of the red seaweeds. The Foliaceous Coralline (*Membranipora pilosa*) runs in very narrow ribbons, covered with a "pile" of bristles, up the stems of various weeds; and many another of the nearly two hundred and fifty British species will be sure to fall to the patient and sharp-eyed investigator.

The horny cell in which the polypide resides is really its own cuticle or outer skin, to which it is inseparably attached. If careful examination be made, it will be found that at the mouth of the so-called cell the horny material suddenly changes its character and becomes a very fine and delicate tissue, capable of the greatest freedom of movement and folding such as is absolutely impossible with the horny portion. This remarkable change of character in the two portions of the same cuticle allows the anterior portion of the polypide, with its crown of tentacles, to be suddenly and completely withdrawn out of danger, just as easily as the tip of a glove-finger can be withdrawn into its lower portion.

The tentacles that encircle the mouth of the polypide are hollow, and covered with ever-waving cilia, whose beating

causes currents of water to set in towards the animal's mouth,
bringing food with them. These tentacles appear to be also
the only sense organs possessed by the polypide, and to serve
the further purpose of gills. None of the Polyzoa to which we
here make reference possesses a heart or blood-vessels.

CHAPTER V.

JELLY FISHES.

IT has been remarked that we get our best ideas of geography from the newspaper-man's special correspondence in war time. Certainly, at such times certain places that are not even marked on ordinary maps are thrust into such prominence that they become familiar to thousands who otherwise would never have known of their existence. In a similar fashion many scraps and fragments of useful knowledge that will stick in the memory will be picked up by the newspaper reader who is simply bent on following the moves in the great political game. For instance, it is not many years since a well-known Scots peer, in order to cast ridicule upon his opponents, enlightened the world upon the subject of Jelly-fish organization. The party he held up to scorn resembled Jelly-fishes in his estimation because they were invertebrate—they possessed no backbone, and could make no progress against the tide, but were forced to float aimlessly with the current. The political small-fry took up the parable from the venerable duke, some reproducing it with variations that appeared marvellous indeed to the mere naturalist; but it was soon quite generally known without recourse to text-books, that the Jelly-fish was not a vertebrate animal, and that it had no muscular power sufficient to enable it to move against the tide.

Now these facts in the natural history of the *Medusæ*, elementary though they be, are such as in the ordinary way might have taken generations to get fixed on the public mind. Many persons who spend their autumnal holiday at the seaside, become fairly familiar with the more or less broken and lifeless forms of one or two common species, as they get drifted upon the beach and are unable to get off again; but they have probably little idea of the beauty and elegance of these frail creatures when fully expanded and pulsating with life a short distance from the shore.

There are two things which stand in the way of a more familiar knowledge of these Jelly-fish, on the part of the public. First, they are almost entirely composed of water. and, having no muscular tissue, are soft and flabby to the touch—a characteristic which inspires feelings of abhorrence in the average man or woman. A man may courageously face a dangerous wild beast, and yet shrink with loathing and dis-gust from contact with a slug or a Jelly-fish—though, with strange inconsistency, he may swallow a living oyster with gusto! Having found a stranded Jelly-fish on the beach, he will probably turn it over with his stick, call to mind the Duke of Argyll's political simile, and pass on.

The second reason is that certain common forms have an unpleasant trick of stinging slightly. This is a power given to them for the purpose of paralysing small creatures they secure as food, but they have sometimes mistakenly exerted it upon a timorous thin-skinned bather, against whom they have drifted.

There are, however, only two or three of our native species that have that power, and though they have been known from ancient days as Sea-nettles, Stingers, and Stangers, there is no doubt that their virulence has been greatly exaggerated. This exaggeration probably owes something to the graphic word-picture of the late Professor Forbes, in which he des-cribed the Hairy Stinger (*Cyanea capillata*). In picturesque language he depicted it as " a most formidable creature, and the terror of tender-skinned bathers. With its broad, tawny, festooned and scalloped disk, often a full foot or more across, it flaps its way through the yielding waters, and drags after it a long train of riband-like arms, and seemingly interminable tails, marking its course, when the body is far away from us. Once tangled in its trailing 'hair,' the unfortunate, who has recklessly ventured across the monster's path, soon writhes in prickly torture. Every struggle but binds the poisonous threads more firmly round his body, and then there is no escape, for when the winder of the fatal net finds his course impeded by the terrified human wrestling in his coils, seeking no combat with the mightier biped, he casts loose his envenomed arms and swims away. The amputated weapons, severed from their parent body, vent vengeance on the cause of their destruction, and sting as fiercely as if their original proprietor gave the word of attack."

No doubt Forbes had good grounds for his statement in the experience of one of these delicate and nervous persons who suffer more mentally than physically, and whose imaginative powers would create a horror out of their contact with a spider, or even its web. The mischief is that the bookmakers, who have no practical knowledge of their subjects, go on quoting Forbes approvingly, and on this slight foundation characterise the whole jelly-fish race as stinging creatures. It seems very probable that some of the larger tropical forms that have the stinging power are far more virulent than those inhabiting British seas; but I have handled the Hairy Stinger and lifted it from the water with my bare hands and experienced no discomfort from the operation.

The Rev. J. G. Wood improved upon Forbes, and described the pain inflicted by *Cyanea* as being at first like that following contact with the stinging nettle of our hedgerows; getting more severe it causes a sharp pain to flit right through the nervous system, the heart and lungs suffer spasmodically. This state of affairs lasts for ten or twelve hours, and then for several days the skin is so sensitive that the sufferer can scarcely bear the contact of clothes; and it is months before the shooting pains depart.

With such a character it is little wonder that the unscientific public should decline an intimate acquaintance with the
family. And yet the story they have to tell is as marvellous
as any that will be found in the whole range of Mr. Lang's
Blue, Red, and Green Fairy Books. It is the story of the
insignificant and despised dwarf, who one day bursts through
his squalid exterior and stands revealed as the handsome
prince magnificently attired, whom all the princesses desire to
marry. It begins in the orthodox way with, Once upon a time
there was a simple and very tiny creature, with soft white flesh
and no bones, who dwelt on a rock on the sea-shore. He was
just a little tube of jelly, and though he had a mouth he had
no head. His many arms were arranged in a circle round his
mouth, and from his body sprouted out several creatures like
himself, but much smaller. Learned men had examined him
and declared that his proper name was *Hydra tuba*. He
remained fixed to this rock from the autumn right through the
winter's storms, and in the spring it was noticed that he was
getting old, for a large number of wrinkles appeared on his
tubular body. Weeks went by and the wrinkles became
deeper and the edges of them turned up, so that the upper
part of the creature's body looked like a dozen saucers piled
up one in the other. Then these saucers each grew a series
of eight arms from its edge, and the uppermost of the pile
broke away from the others and began to float off through the
water. The next, and the next, and every one of the remaining saucers floated off in the same fashion, and those who
watched them do so, say that they gradually grew into glorious Marigolds or Sea-Jellies, with umbrella-like bodies of clear
jelly, marked on the top with rings and streaks of red, and all
around its edge each had a delicate fringe looking like the
finest of silk. And so they floated off to see the world and
seek their fortunes.

The Jelly-fish produces ova, which develop *cilia*—eye-lash-
like processes, by means of which they swim through the water.

Settling on a rock or shell, they develop into *Hydra tuba*, with long tentacles, as at *a a*. Then comes the saucer-like stage, as at *b*; finally the free-swimming segment, *c*, which ultimately becomes the huge creature of our next illustration, which is so plentiful in our seas during summer and early autumn.

LARVÆ OF AURELIA.

Every person that has any acquaintance with Jelly-fishes at all knows this species well—by sight. It is probable that many of those who think they know it would be somewhat puzzled if asked to point out the creature's mouth and to give a rough outline of its organization. It may be described roughly as umbrella-shaped. There is an arched disk, from the centre of which, on the concave or lower surface there depends a thick cylindrical body, the *manubrium* or handle, sometimes erroneously termed the *polypite*, which finally terminates in four lobes assuming the form of trailing ribbons. In the centre of these lobes is the creature's mouth, and the stomach is continued from the mouth *up* the middle of the manubrium. Here digestion takes place, and the nutriment thus obtained is carried up to the centre of the umbrella, and thence distributed to all parts by means of nutrient tubes which may be seen running straight from the centre to the circumference. Looked at from above, the *Aurelia* will be seen to have its disk symmetrically marked off into eight portions by these nutrient tubes, each of which reaches the edge where there is a little notch, and then continues round the margin. Now at the notch there is a ganglion, or nerve centre, a kind of local brain, for the Jelly-fish is very low in the scale of nervous

organization, and possesses no central brain; in fact, its
ganglia are only the beginnings of a nervous system of primitive type. At one time these spots were thought to be eyes,
and the Jelly-fishes were divided into naked-eyed and hooded-
eyed according to whether these sense organs were covered

MARIGOLD (*Aurelia aurita*).

with a kind of flap or not. It is now more clearly established
that they are olfactory organs, possibly in some cases they
combine the functions of both nose and eyes. They are
known to naturalists as *tentaculocysts*. The *Aurelia* moves
slowly through the water by the alternate expansion and contraction of its umbrella-disk. The four crimson lunar marks
on the disk are the ovaries in which the eggs of the Jelly-fish

are produced. The eggs make their way through the stomach to the mouth of the manubrium, where there are little cavities for their reception, and here they stay until they have developed a fringe of cilia, when they swim off. In this condition they are quite flat, and of old they were regarded as a distinct species of animal under the name of *Planula*. It afterwards becomes pear-shaped, tires of wandering, and settles down on a rock or shell to undergo the series of developments we have already described, every stage of which was formerly considered a different animal and bore its special name.

In spite of the structureless appearance presented by these Jellies—owing to the presence of a thick layer of transparent gelatinous material—they are endowed with true muscular fibres, which are confined to the under surface of the umbrella, to the manubrium and tentacles, and to a flap of the umbrella margin which is directed inwards and known as the *velum*. It is by the contraction of the velum that water is expelled from beneath, and this has the effect of forcing the Jelly-fish in the opposite direction.

Somewhat similar to the *Aurelia* in general form is the Hairy Stinger (*Cyanœa capillata*), to which allusion has already been made. Its umbrella is not so disk-like, but has a raised central dome, and its edges are beautifully fringed with long threads. The lobes around the mouth are developed into very long appendages, all frills and furbelows. An allied species, *Cyanœa chrysaora*, has a very thick and bulging manubrium, but no long streamers depending from it.

A very common form which swarms in harbours is *Thaumantias*, of which there are several species. In these the jelly is very thick at the crown of the umbrella, which is more bell-shaped than in *Aurelia* or *Cyanœa*. The nutrient tubes are four, and the ovaries are beside them. A very small species, *Turris digitalis*, is bell-shaped, with a conical top and a deep fringe of tentacles round the margin. It originates as a polypite on a so-called coralline similar to those described in Chapter IV.

On our South-western shores we sometimes receive visits
from Jelly-fishes which must be regarded as distinguished
foreigners.   Among these is the beautiful creature to which
seamen give the name of the "Portuguese Man o' War"
(*Physalia pelagica*).   It is of a shape entirely different from
those we have noted.   Instead of an umbrella it has a spindle-
shaped bladder distended with air and coloured with blue,

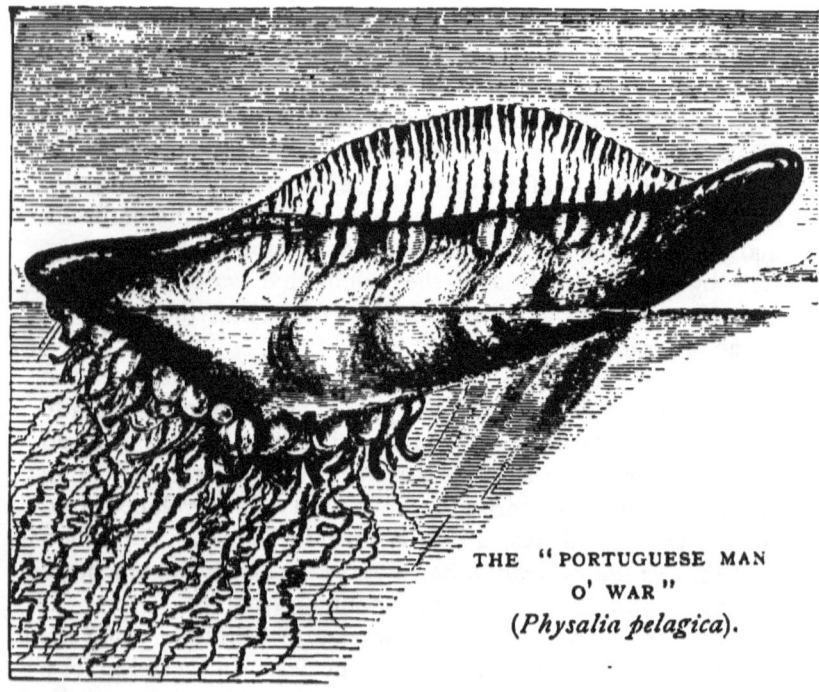

THE "PORTUGUESE MAN
O' WAR"
(*Physalia pelagica*).

whilst along its upper surface there runs a beautiful pink frill
which serves as a sail.   From the lower surface there hangs
down a cluster of long trailing corkscrews, beautifully coloured
and capable of stinging.   Occasionally individuals from this
floating colony develop into Jelly-fish of distinct form, and
swim away from the community.

There seems no doubt of the stinging powers of this species, for Dr. Bennett, a naturalist, has given us his account of the unpleasant effects following upon his handling of this " Man o' War." He took hold of the bladder, and the creature raised its long appendages, twining them round his hands and stinging with great severity, and clinging so tightly that he had difficulty in removing them. He says the pain was like that caused by severe rheumatism, and extended up his arm to the muscles of his chest. Symptoms of fever followed, with rapid

TUBE-MOUTHED SARSIA.

pulse and difficult breathing. This continued for three-quarters of an hour; but even then he was not free, for his skin was marked with raised white wheals for several hours. The tentacles, he says, can be thrown out to a distance even of eighteen feet for the purpose of stinging its prey. This species is often met by mariners in extensive fleets, so to speak, and sometimes great numbers of them are wrecked upon the coasts of Devon and Cornwall; occasionally they have been found on the eastern shores of England, but they really belong to the Mediterranean and the open ocean.

A common form on the south coasts is the Tube-mouthed Sarsia (*Sarsia tubulosa*), of which we give a portrait. It is bell-shaped, with what looks like a very long clapper hanging from its centre, and four long tentacles from the edge. The clapper is, of course, the manubrium, and contains the mouth and stomach, which it can stretch out to very accommodating proportions. The bell is only about half an inch in height, but the manubrium is more than twice that length.

FORBES' ÆQUOREA.

Scarcely to be found on our coasts away from Devon and Cornwall are two species of *Æquorea*, of which the one repre-sented is dedicated to the memory of the late Professor Edward Forbes, who did so much to extend the knowledge of marine life, especially in relation to jelly-fishes, anemones, star-fishes, and mollusks. It is therefore designated *Æquorea forbesiana*, and a man might well feel proud to have so beautiful a creature named with his name. It is a little larger than our figure. Its upper portion is of thick crystalline jelly, coloured with a lovely sky-blue tint lower down. Below the blue region are a number of curved lines of bright crimson—the nutrient channels—and the four lobes of the manubrium are similarly coloured. There are streaming tentacles around the margin which lay hold of minute creatures that pass by. The early history of this form of Jelly-fish is unknown—whether it passes through stages resembling those of *Aurelia* or of *Turris*, or attains the medusa-form direct from the egg. Any of our readers that may have the opportunity for observing this beautiful creature, should make a point of recording what he sees. It may be of great assistance in working out the true relation of this species to other forms.

One that must be classed with the " Portuguese Man o' War" as a visitor to our south-western coasts, is called the Sallee Man (*Velella scaphoidea*), a kind of Jelly-raft, upon which is hoisted a little sail, and whose margin is fringed with tentacles. As in *Physalia*, the under side of this float consists of a colony of many individuals, which from time to time develop into free-swimming jellies.

But in spite of the colour-glories and imposing size of these larger forms, we have upon our shores swarms of a veritable gem that, in its way, for delicate beauty outshines them all. It is the Globe Beröe (*Pleurobrachia pileus*), sometimes called the Sea Gooseberry. In early summer, when the seas are still, and everything for five fathoms or more can be clearly seen through the crystal waters of the Cornish coast, this fairy form may be clearly seen in spite of its short diameter (half-inch) and its perfect transparency. You are lazily drifting in a boat, but your eye catches minute flashes of iridescent colour in the water, and you must lean over the boat's side to see what it is. You then discover a number of these crystal globes passing gracefully and without seeming effort through the water, not always in our plane, now upwards to the surface, then downwards out of reach. You are fascinated by the exquisite beauty, and hope the one you are watching will not pass out of sight. As if in response to your unexpressed wish, it ceases its downward course, and whilst suspending itself in the water begins to revolve laterally. Your astonishment increases, for you now see that it is furnished with paddle-wheels, or else it is an animated paddle-wheel itself. No; a turn convinces you again it is globular in form, but the paddles are equally obvious. How can it be? What machinery turns them? and what are the two almost interminable threads of gossamer that trail behind, below, or above it?

You watch your opportunity, and the next time one comes near the surface you skilfully trap it in a glass jar, and then some of its mystery is made clearer. The paddle-wheels are eight bands that stretch from pole to pole, and across these at short intervals are rows of " eye-lashes." There is no rotatory motion of the bands, but by the alternate depression and raising of the eye-lashes, an optical illusion is produced. And yet the effect is the same as if these bands revolved with fixed " floats; " the movements of the eye-lashes *row* the fragile vessel through the water, and with every movement the light is reflected in prismatic tints that seem to pass in rapid flashes along each of the eight bands. But whilst we have been investigating the mystery of propulsion, what has become of those long attenuated streamers? Broken off by our rough handling as we potted it? No; the creature, as though sensible of danger, has carefully tucked them away into suitable pockets. We can behold them through the clear jelly in curved club-shaped receptacles. Look; here they come! The Beröe is getting confident, and the tentacles stream out again to six times the length of the animated globe. They can be lengthened or shortened at the creature's will; and each one is provided with an enormous number of short side-branches, like tendrils on a vine. There is no pendulous stomach and mouth hanging from the floating body, for the Beröe differs from the Marigolds and Stingers, and is more closely allied with the Sea-Anemones. Its mouth is at the top of the globe, and its digestive cavity is central.

I have described its appearance as seen from a boat, but it must not be inferred that it cannot be obtained from the shore. A sharp eye will see them in ports and harbours when gazing from low rocks or landing slips. If our reader is desirous of watching these, a few should be entrapped into a clear glass jar of sea-water, but other creatures should not be introduced. I find that small crabs, prawns, or even anemones are not to be trusted with *Pleurobrachia*, or these will rapidly disappear.

The creatures we have brought together in this chapter under the popular term Jelly-fish, really belong to very distinct groups of animal

BERÖE AND YOUNG.

life, and their developmental histories are different. Many of them, in fact, are nothing more than buds from the branching Zoophytes incorrectly called corallines that grow from shells and stones, and of which we have had something to say in a previous chapter.

Agassiz has described a huge form of Stinger (*Cyanæa*

arctica), with an umbrella-disk like those we have mentioned, but measuring no less than seven feet across, yet originating as a bud from a lowly coralline not exceeding half-an-inch in stature. Others are not solitary individuals, but companies of polyps that share the organ which bears them through the waters. Such is the case with the *Physalia* and the *Velella*, the appendages of which consist not of one mouth and stomach, but of many.

I remarked near the beginning of this chapter that the Jelly-fish was very largely composed of water. Professor Owen, not content with having to make an indefinite statement of that kind, went carefully into the matter of pounds and ounces and grains. He said: " Let this fluid part of a Medusa (Jelly-fish), which may weigh two pounds when recently removed from the sea, drain from the solid parts of the body, and these, when dried, will be represented by a thin film of membrane, not exceeding thirty grains in weight."

As a practical illustration of the value of having *that* amount of knowledge respecting such trivial things as Jelly fish, the late Robert Patterson, F.R.S., gives the following story, which was told to him as a personal experience by an eminent zoolo-gist, whose name he does not mention.

"This gentleman had been delivering some zoological lectures in a seaport town in Scotland, in the course of which he had adverted to some of the most remarkable points in the economy of the Acalephæ. After the lecture a farmer, who had been present, came forward and inquired if he had under-stood him correctly, as having stated that the Medusæ con-tained so little of solid material, that they might be regarded as little else than a mass of animated sea-water. On being answered in the affirmative, he remarked that it would have saved him many a pound had he known that sooner, for he had been in the habit of employing his men and horses carting away large quantities of Jelly-fish from the shore, and using them as manure on his farm; and he now believed they could have been of little more real use than an equal load of sea-water. Assuming that so much as one ton weight of Medusæ, recently thrown on the beach, had been carted away in one load, it will be found that, according to the experiments of Professor Owen, the entire quantity of solid material would be only about four pounds of avoirdupois weight, an amount of solid material which, if compressed, the farmer might, with ease, have carried home in one of his coat pockets."

Let me, in closing this very inadequate glimpse of a most interesting group, add that many of these creatures contribute to that phosphorescent appearance of the sea, which is such a wonder and a revelation to those who behold it for the first time. The limits set for the entire volume will not permit me to deal with all the British species; but I trust sufficient has been said to awaken real interest in these despised Stingers, Jellies, and Sea-blubbers.

CHAPTER VI.

SEA-ANEMONES.

THE visitor to a rocky coast possesses the greatest advantage for the study of the Sea-Anemones. These are among the surprises for the inlander whose get-up is not too fine to allow him to scramble over the rocks. If he has already gained some introduction to the beauties of form and colouring in this group, and wishes to get a more intimate knowledge, let him visit some coast village in South Cornwall.

Anemones, with few exceptions, dislike a muddy shore, and are not very partial to sand; nor are they easily seen in thick water. But where the cliffs and fringing rocks are hard and insoluble, the waters are crystalline, and every detail of life in the rock-basins, and even on the submerged reefs, can be plainly observed. Such conditions Cornwall offers, and there anemone-life may be said to attain its greatest luxuriance.

Between the limits of high and low-water the rocks will be found thickly studded with the common Beadlet (*Actinia equina*), in several well-defined colour varieties. In the rock-pools more Beadlets, with a few large specimens of the Opelet (*Anemonia sulcata*) and many young ones. Huge Dahlia Wartlets (*Urticina felina*) lurk under gravel at the bottom. Almost invisible, though exceedingly abundant, are the Daisies (*Cereus pedunculatus*) and the Gem Pimplets (*Bunodes verrucosa*).

For others we must wait until the ebb of the spring-tides brings us a few days of exceptionally low water. Then, when we can get to a floor of a big "drang," like that shown in our frontispiece, we may take such species as the Plumelet

(*Metridium senilis*), the Rosy Anemone (*Sagartia rosea*), the Snake-locked (*Cylista viduata*), the Globehorn (*Corynactis viridis*), and others to be named hereafter.

Now these, I think, make a fairly representative list, though it by no means exhausts the British species. However, if the reader can manage, during two or three visits to the seaside, to find and identify the species named, he will be fairly well acquainted with the Anemones of the shore, as distinguished

THE BEADLET.

from those found solely in deep water, which can only be explored with the trawl or dredge. With this class we have no concern in the present volume, as the deep water is beyond our province; except so far as certain of them may come ashore attached to deep-water mollusks and crabs.

The Beadlet being at once so widely distributed round our shores, and so abundant wherever found, becomes very suitable for use as a type of the whole class. We are not to be

tempted into repeating or plagiarising the gushy nonsense that
has been so lavishly poured out by many writers, in which the
Anemones have been commended to popular notice because
of their wonderful resemblances to flowers.   Even the older
naturalists were not free from blame in this matter, for they
named the animals zoophytes (animal plants) and anthozoa
(flower-animals), names that have stuck, and of which we
cannot be rid.   The term "anemone" (wind-flower) itself is
utterly absurd when applied to the Actinia.   Beyond the
brilliant colours and the petal-like rays of certain species,
there is no parallel between these creatures and flowers, and
the institution of such poetical similes in too many cases only
serves to hide the true nature of these interesting forms of
life.

On a rocky coast at low water we shall find the Beadlet
thickly studding the rocks that stand up high above the sand
or pebbles.   Those that are on the perpendicular face of the
rock are smooth hemispheres of dark-crimson, bottle-green,
olive, or ruddy-brown, with a more or less vivid thin margin of
blue where the base is attached to the rock.   Lower down, a
little above the water, we shall find them more elongated and
hanging downwards, some with the rays or tentacles partly
extended, but the whole animal looking somewhat flaccid.   *In*
the water, however, whether it be of the rock-pool or the
actual sea, the tentacles are so widely spread that, looking
down upon them, we can see but little of the fleshy column or
even of its base.   These tentacles are never very long in this
species, but they are fairly numerous, there being 192 in an
adult specimen, arranged in six series.   Their general tendency is to arch over towards the column, and so hide the
row of blue eye-like spherules that peep out between the
column and the tentacles.   Within the radius of the tentacles
is an almost flat, smooth expansion of flesh, called the disk, in
the centre of which, on a conical eminence, is the mouth.
The mouth is the opening of a bottomless sack which serves

as stomach, and from the internal cavity, into which the digested food falls, there are channels which convey it all over the body, whilst the indigestible portion is rolled up and thrown out by the way it entered.

The entire quantity of solid matter in an Anemone is very small, as may be seen in certain species (*i.e.*, the Snake-locked Anemone) that become exceedingly thin and flat in the day-time, but expand into a tall graceful column at night. In a similar fashion the tentacles are constantly withdrawn by becoming very small; and the full expansion of these and of the column is alike affected by the absorption of much water.

Most of the Anemones attach themselves to rocks, shells, or weeds, by means of the broad base of the column; others have a rounded base which is thrust down into sand and there retained by inflation. They can move on this base, much after the manner of a snail or slug, but more slowly; some, such as the Opelet, constantly inflate it to such an extent that it becomes a swimming bladder, buoying them to the surface of the water, along which they float inverted.

Reproduction takes place in three ways: first, a division may take place across the disk and mouth, and this be continued right down the column to the base; second, buds may appear on the disk or column and develop into complete Anemones; third, by eggs, which are usually retained until the germs have developed a row of tentacles, when they are cast out from the mouth in batches. This last is the commonest mode; and the extruded young at once attach themselves to the surface upon which they fall.

The Beadlet gets its popular name from the row of blue bead-like spherules to which notice has already been directed. In one well-marked variety of this species the spherules lose their azure hue and become quite white, whilst the normally blue line at the base becomes flesh-coloured, or is entirely absent. There are many other colour variations which it would be foreign to the purpose of a simple handbook to

enumerate in detail.   We will mention one, because otherwise
it might be taken for some other species : if the ground colour
of its column is green, it may be marked with short lines or
dashes or spots of yellow ; or if it is dark-red or liver-coloured,
it may be studded with green dots.   It is one of the hardiest
kinds to keep in an aquarium, where it will soon multiply by
discharging a number of tiny replicas of itself, though sometimes these will be sent out as mere eggs, which will not get
their tentacles until a week or ten days later.

There are several species of Anemone which, though they
differ strongly in the eyes of a naturalist, may easily be confused with the Beadlet on a cursory glance when they are in
the " button " or closed condition.   Two of these are represented in this illustration.   The Rosy Anemone (*Sagartia*

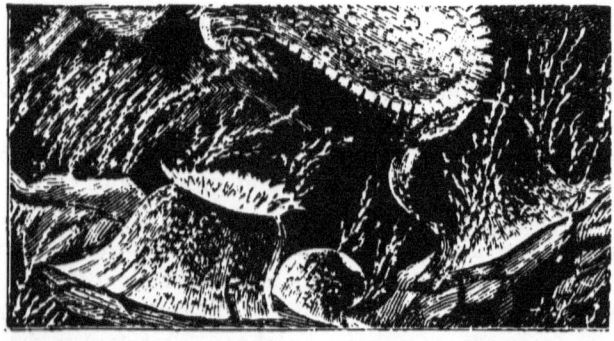

SNOWY ANEMONE.   ROSY ANEMONE.

rosea) is representative of an entirely different genus from that
to which the Beadlet belongs.   When expanded the column is
cylindric in shape, its base not nearly so broad as that of the
Beadlet.   Near the base the colour is buff, deepening above
into a rich ruddy brown ; on the upper part there are a number of little suckers, to which fragments of shell and gravel
adhere.   The tentacles are rosy, with an inclination to become

purplish, and some of them are indistinctly marked by two transverse bands of a darker hue.   The disk is pale olive, and the mouth white or pinkish white, not raised like that of the Beadlet.   Its usual habitat is in rock-pools that are uncovered only at very low water.   One called the Pallid Anemone (*S. pallida*) is, I feel sure, a mere colourless variety of *S. rosea*.

ORANGE-DISK ANEMONE.

The Snowy Anemone (*Sagartia nivea*) is in form much like the last-mentioned species, but its column is coloured pale olive-brown, paler near the base.   The whitish suckers on the upper part are more prominent than in the Rosy Anemone. The disk, the tentacles, and the mouth are all a beautiful white.   It will be found in the low-lying rock-pools.

There is another species, nearly allied, that has white tentacles with grey tips, but the disk is of a dull, orange tint, with a dusky border at the roots of the tentacles.   This is the Orange-disk Anemone (*Sagartia venusta*), a species that likes to hide its brown column in a hole or a crevice of some over-hanging rock or in a cavern.   It is very local.

S. venusta is very shy, and readily folds in her tentacles— in truth, she seldom opens them very widely.   *S. rosea*, on the other hand, will fully display her charms immediately after she has been transferred to an artificial home.   *S. venusta* is but little inclined to rove about an aquarium, but whenever she does so she appears bound to leave a portion of her base behind her.   In the course of about ten days this detached portion develops tentacles, and sets up an independent existence.

There is a group of which the members are invisible unless their tentacles are expanded, and even then they harmonise so well with their surroundings that they are seen only with difficulty.   Of course, when the eye has got accustomed to their forms and colours, and knows what to look for, it finds them, if they are present.

I have shown a small rock-basin to a friend whose eye is pretty keen where natural objects are concerned, but he has utterly failed to see the crowd of Anemones in full expansion that were there, until several had been almost touched by my finger in pointing them out; then a minute or two later he was finding out the others that were there, without any assist-ance from me.   One of the species concerned was—

The Cave-dweller (*Cylista undata*), which is exceedingly liable to variation in form and colour.   It is difficult to obtain, owing to its awkward habit of fixing its base in some narrow chink of the rock, and spreading out its broad disk above the crevice.   Unless one is very careful in excavating the troglo-dyte, one may cut it in two, or hopelessly smash it.   The column is of a dirty-looking drab colour, shading off into grey near the summit. The disk is variable in colour, and indefinite in detail; but in general effect it is a minute patchwork of black, brown, and yellowish-drab lines radiating from the whitish mouth, and minutely dotted with white. From each angle of the mouth there is a very distinct, short, opaque white line. The tentacles are numerous, and variable in length; their ground colour is a clear grey, with several cross bands of white, and at the base there are two small patches of white surrounded by black in such fashion as to form an obscure B. This is a pretty constant mark in the identification of the species, though the white patches are sometimes missing.

The Cave-dweller, though not easily seen at first, is widely distributed upon our shores, whether rocky or sandy, and careful examination of the pools on hands and knees will probably reveal large numbers. Occasionally we shall come upon a cleanly-hollowed basin in the rocks, about two feet across and almost as deep, the interior thickly coated with a dense growth of coralline. In this the Cave-dweller and the Gem Pimplet delight to grow, and in such situations they can be more easily obtained than from the chinks of rock that will not admit the fingers.

The Daisy Anemone (*Cereus pedunculatus*) is similar at first sight to the Cave-dweller. Why it was named daisy it is difficult to imagine, for I have seen no specimens that suggested the most remote resemblance to that flower. It is similar to the Cave-dweller both in form and habits, but it is more soberly coloured, the very broad disk being dark-brown or black, crossed by very fine red lines radiating from the mouth and continued along the sides of the tentacles. The brown of the tentacles has a yellowish bias. Near the base of some tentacles there are two white bands separated by a patch of brown; others are uniformly coloured throughout, save for tiny specks of white sprinkled without order over them. The tentacles are very numerous (400 or 500), and mostly small. The Daisy is found in the crevices of pools left by the ebbing tide rather than in those of perpendicular rock-walls.

The Scarlet-fringed Anemone (*Sagartia minia'a*) has a crimson-brown column with buff-coloured suckers. The disk is greenish grey with darker mottlings. The tentacles are clear glassy, of a brown tint with darker rings; on the surface a pair of longitudinal dark lines converging to one point at the tip of the tentacle; at the base two patches each of black and white alternating. The outer row of tentacles differ by having the interior coloured with orange or scarlet, which shows clearly through the thick but colourless substance of the tentacle; from these the Anemone gets its name of Scarlet-fringed. It inhabits holes in the rock-pools, and in the rocks of deep water, but does not affect such deep and narrow crevices as the Cave-dweller.

Occasionally, when we are engaged in stone-turning at low-water, we shall come across a colony of the pale spectral forms of the Snake-locked Anemone (*Cylista viduata*), but it is one found only with difficulty, because in the daytime it compresses itself into a dirty yellow button as thick as about six of the pages of this book, with pale lines radiating from the centre. In this condition it offers to the eye the appearance of a limpet-shell, or a flake of rock. I once found a colony of seven individuals on the back of the Gabrick Spider-crab (*Maia squinado*), where no doubt they had been planted by the crab with a view to getting artistic effects. This suggests to us that some of the deep-water species, not referred to here, may be obtained by examining the shells of oysters, quins, whelks, etc., which are dredged in deep waters.

Supposing you have been fortunate enough to find a small-sized stone supporting two or three of these compressed Anemones (*Cylista*), and having taken it home have placed it in a thin glass tumbler of sea-water for observation. At night you look at the glass to see how the strangers are doing, and behold with astonishment the change that has taken place. The depressed yellow button has gone, and where it lay there stands a tall and elegantly-formed column, two inches in height, tapering from the base and the summit to the middle, and supporting a crown of many pellucid tentacles. The inner row of these stand up and arch outward, the outer ones hang out a little way and then droop with perfect grace. The contrast between the two conditions is really startling; and as you observe the tentacles slowly but continuously writhing you admit the propriety of the English name.

THE OPELET.

The column is marked with a series of paler longitudinal lines, and on its upper portion there are small suckers, though the creature does not appear to use them. The distinctive mark of the species is its long, lithe, transparent grey tentacles. There is a very fine black line running along each side of these, and at right angles to them is a couple of bands of white—one at the base and one about the middle.

On the rocks that are uncovered only at the recess of the spring tides, and in the shallow pools a little higher up the shore, we shall find abundant supplies of Opelets (*Anemonia sulcata*), here buried in holes with only the tentacles protruding, there attached to the bare rock-surface and exhibiting a substantial brown column, short but very broad, and bearing an innumerable, almost disorderly crowd of snaky-tentacles, ever writhing and intertwining. In some specimens these are a lovely lustrous green with lilac tips; in others grey, or white, or yellow. The grey and the green are the most abundant forms, and we may take the satiny green as the typical form. One peculiarity will soon strike him that makes its acquaintance for the first time—that unlike the other Anemones he knows, he cannot see one with tentacles withdrawn. There is no button stage in the Opelet, but there is a corresponding restful condition when the waters have receded from its rock, and the previously solid-looking column has collapsed, and the flaccid tentacles hang in an empty, lifeless manner among the weeds. The Opelet does not settle down permanently on one spot. He likes a change, and so never attaches his broad base very strongly. It is easy to get him off the rock when he is wanted for an aquarium specimen, and it is equally easy for him to slide off, and, inflating his base to a great size, float on the surface of the water with his tentacles waving downwards.

The Opelet attains a great size, and then appears to delight in sitting on the broad leathery fronds of *Laminaria*, with which his olive column harmonises well.

I had a specimen for about eight months that practically filled a bell-glass, nine inches in diameter. Stationed in the middle, he could nearly touch the glass all round with the tips of his tentacles; as a matter of fact he was nearly an inch away, which meant that the area occupied by his tentacles was at least seven inches across, and when he chose to inflate himself fully he could improve upon this. He was a very voracious feeder, and there was always room in his capacious column for a good meal. Alas! he was a victim to gluttony. One day I brought home a Butter-fish, or Gunnel (*Murænoides guttata*), about six and a half inches in length. Thinking he was large enough to take care of himself, I put him in with the big Opelet. He had been there but a few minutes, when I looked in to see how he was settling down in this new world. He was already dead or insensible, in the snake-like folds of the green tentacles which were tightly coiled around the fish. I attempted a rescue, but these tentacles are wonderfully adhesive, and feel as though they had been painted with patent glue: they adhere on the slightest touch.

I was too late to save his life, so I did not trouble to recover the corpse. Before long it had reached the mouth, which extended considerably in order to accommodate it; but it was a little while before the intelligence of the Opelet could be so brought to bear on the matter in hand that the Anemone could comfortably get the Gunnel "end on." Now the task was easy, and although the Gunnel considerably exceeded the Opelet in length, the Anemone tucked him safely in. It was not a comfortable arrangement in spite of the elasticity of the Opelet; and the fish, as could plainly be seen from outside, had to be slanted. Whether this caused a rupture of any vital part, or whether the Gunnel was too much for the Opelet's digestive powers, cannot now be ascertained; but the Opelet sickened, and though the fish was discharged next day, the Anemone never recovered, but finally died about a week after this inordinate meal.

The late Mr. Gosse experimented upon the Opelet as an addition to our breakfast table, and declared it good. He says that "the dish called *Rastegna*, which is a great favourite in Provence, is mainly prepared from the Opelet."

Perhaps some of our readers would like to experiment in the same direction whilst they are at the seaside; in that case we should be glad to have their experience and candid opinion on the suitability of our native Anemones for human food.

Dr. Andrew Wilson, in the days of his youth, desirous of emulating Mr. Gosse's example, cooked a specimen of the Dahlia Wartlet, but the result was not such as to confirm him in this line of alimentation, though he admits that the Dahlia is probably a tougher subject than the Opelet, and requires different treatment to make it equally inviting as a *bonne bouche.*

One of the most delicately beautiful of our Anemones is the Gem Pimplet (*Bunodes verrucosa*), which may be sought in rock-pools near low-water; also at low-water, half buried in the sand, at the base of rocks.

Its name of Pimplet is a soft way of describing its column, which is crowded with pimples. As a rule these are of a light pinky-brown or rosy tint, diversified by six vertical bands of larger white pimples. In several specimens I have before me as I write, however, the column is uniformly grey with a pinkish tinge, the pimples being of the same hue and of equal size. The disk is dark-grey, marked with fine lines of the darker rays proceeding to the tentacles, and the space around the elevated mouth is yellow, marked with a small clear spot of carmine at the angles of the lips. The tentacles are conical, rounded, with blunt tips; the underside transparent grey, the upper side darker, with many thin lines and broad rounded bars of opaque white across it. When the tentacles are withdrawn and we have the rounded top of the button stage, the effect of the six white rows of pimples converging at the summit and forming a star pattern is very pretty. But when the whole of the tentacles are fully expanded, the outer row bending slightly downwards, the next row curving upwards and outwards, whilst the inner ones stand more or less erect, the effect of the delicate pencillings and the pellucid greys in contrast with the warmer tints of the column is exceedingly fine.

When the specimens are growing in a coralline-lined basin, however, this peculiar style of beauty does not render them at all conspicuous; on the contrary, the Gem Pimplet is a species that will not fall to the hasty collector who rushes with a mere glance from pool to pool, but it will soon reward the careful and patient investigator who is willing to recline at the side of a small pool until his eyes have closely scrutinised every inch of the bottom, and given the fixed objects a chance of revealing themselves by a slight movement.

Owing to the transparency of the tentacles in *B. verrucosa*, an interesting point in the natural history of the species may be observed without difficulty. The larvæ are retained by the Pimplet until they have developed their first series of tentacles, and the hollow tentacles of the parent are made use of as convenient receptacles in which to store the brood until it is ready to be sent forth into the surrounding waters. Four or five of these may be seen in one tentacle. For some time after their discharge these young Pimplets are exceedingly beautiful. They are pellucid, and in them the remarkable structure of anemones may be clearly seen. When first excluded they are nearly globular, about one-twelfth of an inch in diameter, crowned with a double circlet of tentacles, the outer arching outward and downward, the inner more erect. Within a few minutes they have increased in size to one-sixth of an inch, by the mere absorption of water, their tissues becoming relatively more transparent, and their forms protean. From the globular form they have quickly changed to one more cylindrical, or to a cylinder with a bulbous base, then to a long inverted cone.

The Pimplet is easily removed; he has not got that unpleasant habit of squeezing himself into a crevice, like the Cave-dweller; and when placed in the aquarium he shows no resentment of his change of quarters, but makes himself at home and reveals his beauties at once, even before he has well fixed his base.

An allied species, the Red-specked Pimplet (*Bunodes ballii*), may be found under stones at low-water, but is more frequent in the deeper water outside our zone. It is of a warmer hue

than the Gem, its pimples less prominent, and each one with
a tiny crimson speck at its centre; the interspaces between
the pimples being freckled with crimson. In the aquarium it
will be found to select an obscure angle between the floor of
the tank and a stone. It is very sluggish, and readily settles
down to aquarium life.

In strong contrast to the quiet loveliness of the little Pimplet,
is the massive and showy beauty of the Dahlia Wartlet (*Urticina felina*). The Pimplet reaches up to the light and adds
grace to its beauty; but the Dahlia Wartlet spreads itself out
as widely as possible, so that its diameter exceeds its height
about three times. In spite of its size and its magnificence,
one has got to learn *how* to see it before it appears at all
plentiful; *then*, if we are on the rocks near low-water, we shall
find it in abundance. It is fond of crevices and places where
gravel and broken shell accumulate. Beneath these it buries
its broad base and attaches bits of shell and stone to the many
whitish suckers with which the upper part of its dark crimson
column is thickly studded, and when the tide recedes and
leaves it, the collector has to look, not for an expanse of
brilliant tentacles, but for a little rounded heap of gravel. In
permanent pools, however, where it has crimson weeds and
white corallines around to harmonise with its bright hues, the
Wartlet seldom closes, except for the purpose of securing its
food; there its sucker-warts are little used, and consequently
they dwindle in size. The tentacles are thick, transparent
cones, marked with transverse bands of dark crimson and white.
The disk is of a transparent olive hue at the circumference,
merging into full crimson nearer the centre, where the disk
swells into a low elevation with the mouth in a depression at
its summit. It is a very voracious creature, and its large
mouth and capacious stomach enable it to swallow half-sized
specimens of the Shore-crab (*Carcinus mænas*), sea-urchins,
dog-whelks, and small fishes. On this account it is not so
suitable as an inmate of the aquarium as the others we have

described. It is subject to great variation of colour and markings; that which we have described and figured is perhaps the most plentiful form, but by no means the most beautiful.

There is a pretty little species called the Globehorn (*Corynactis viridis*), to be found by the observant eye, growing in patches on the under surface of overhanging rocks near to low-water, on our south-western coasts. It is seldom a quarter of an inch in stature, and its breadth is a little more; but they are always close together in colonies of from twenty to fifty individuals. It is very variable in colour, but as a rule the members of one colony will resemble each other very closely in this as in other respects. The peculiarity which separates it from the several species we have been describing, is in the form of the tentacles. These, instead of being more or less conical, and ending in a point, consist of globular heads set on stalks —from which circumstance the popular name Globehorn is derived. The column is of even breadth throughout, the base slightly broader, transparent, but coloured white, grey, yellow, green, brown, crimson, or scarlet. Probably the most common form is that which has the column and disk of emerald green. The footstalks of its tentacles are colourless and transparent, but studded with rich brown warts, whilst their heads are rich crimson. The thick-lipped mouth is bright-green.

At low-water we shall probably come upon a rock upon which is a group of dumpy masses of clear white jelly. Carefully remove some of these to your collecting bottles, and in the evening, when they have had time to recover from the shock, they will astonish you. The squat jelly-lump erects itself into a shapely alabaster column, a couple of inches high, and near the top a rounded parapet, above which the lobes of the crown will spread out, densely clothed with feathery tentacles. It is well named the Plumose Anemone (*Metridium senilis*).

In the straightness and tallness of its column, the Plumose Anemone is suggestive of a deep-water species that you may

F

sometimes have brought in shore by a fisherman who has discovered your weakness for what he will term "curios." This is the Parasite Anemone (*Cribrina effœta*), which will almost always be found perched on a full-sized shell of the common whelk (*Buccinum undatum*), or the red-whelk (*Fusus antiquus*). Yet the whelk-shell will not be tenanted by the whelk, but by the Hermit-crab (*Eupagurus bernhardus*). The Parasite, when fully expanded, is about four inches high, and the measurement across the tentacles is not much less. Its column is pale drab in colour, the tentacles creamy white, and the disk somewhat conical. To see a weak creature like the Hermit hauling a heavy-looking shell along is a trifle amusing; but when Cribrina's huge tower of apparently solid flesh is perched on top of that, the picture is absurd. Owing to its large size and its unhappiness when deprived of the society of the Hermit, the Parasite is not a desirable aquarium specimen, except where one has very large tanks affording sufficient depth and range for the Hermit-crab. It is not clear what advantage each of the parties to this strange co-operation gain, though it is easy to propound theories to account for it.

Such partnerships (*commensalism*) are by no means uncommon in Nature; and there is one subsisting on our own coasts between another species of Hermit-crab (*Eupagurus prideaux*) and the Cloaklet-anemone (*Adamsia palliata*). It is probable that the Anemone derives advantage from being carried about from place to place, and thus has better opportunity for securing food than if stationary; whilst the crab is probably saved from being swallowed by a big-mouthed fish, owing to the unpleasant odour of the Anemone. One other way in which the Hermit may benefit is by feeding on the crumbs that fall from the Parasite's table. I have had specimens brought to me that had been hauled up on "spiller lines," the fishermen characterising the Anemone as *an enemy* for stealing his bait. Here probably the advantage gained by being perched

THE DAHLIA WARTLET.

atop of the whelk-shell alone enabled the Parasite to reach and swallow the bait on the spiller hook. It should be added that the base of the Anemone gradually absorbs that portion of the whelk-shell to which it is attached, as may plainly be seen on removing a large individual from the shell.

It may be presumed that a large number of our readers not only desire to be able to identify the natural objects they encounter by the deep sea, but would like, also, to watch the habits and conduct of some of them under more favourable conditions for continuous observation than the constant ebbing and flowing of the tides will allow on the shore. For their benefit let us add a few words.

Certain of the Anemones, which we have already indicated, adapt themselves to the artificial life of an aquarium very readily, and without any great exhibition of shyness. For this purpose it is advisable to take medium-sized specimens, rather than to look out for the largest example we can find, remembering that the younger individuals in time become large; but what is of greater importance they are less likely to be injured by removal. In many cases patient search will show us examples of such species as the Beadlet attached to small stones, and it is much better for our purpose to take these, stone and all, than to disturb the attachment of others to the rock. Others may be found on weeds, especially the broad smooth fronds of the great oar-weed; but some of the more delicate in texture must be removed by chipping off with cold-chisel and hammer a flake of the rock with Anemones attached.

Anemones are not great consumers of oxygen, and consequently the water in the vessels to which they are consigned does not readily become fouled, except as the result of feeding. Do not give food more often than once in a fortnight or ten days, but be sure then that it is suitable food, and in small fragments only. Some people think the best thing to give to such delicate creatures is a piece of raw steak. It is probably

unnecessary to tell *you* that we have the best guarantee
of success when we imitate Nature as closely as possible.
Anemones in a state of Nature do not often get a chance of
raw beef, except when a bullock has been washed overboard
from a ship and comes in a very inflated and "gamey" condition, begging the Coastguard to bury it decently.

If oysters or mussels can be obtained where you are staying, give Anemones tiny pieces uncooked; or a piece out of
the side of a young sole or plaice. Do not give them fish that
is all hard muscle, for they cannot readily digest it. They
require so very little to eat, that we may easily select that
little from a fish that is known to be easily digestible.

Here, too, let me warn you against a misapprehension that
may cause you to be much concerned about the supposed lack
of appetite in your pets. The nutriment they extract from
their food appears to be entirely of a fluid character; they
suck the juices from it, and having done so completely, what
remains becomes pearly white, and having been wrapped in a
thick transparent *glaire*, is thrust out by the way it entered.

Now this excrement is of a very objectionable character,
and if allowed to remain for a short time will infect the whole
of the water in the vessel, and begin to destroy all the life
therein : so it must be removed at once. Persons who have
had no previous experience in keeping Anemones, suppose
that the individual fed had no appetite, and had rejected his
food without change.

The ordinary rectangular aquarium is very suitable for the
reception of the Anemones, and a special piece of rock should
be selected from one of the rock-pools to serve them as a
residence. This stone should not cover more than half the
floor space of the tank; and it should be very irregular as
regards its surface, pitted with holes and recesses into which
the more retiring species may partially withdraw their columns.
If no suitable piece can be found readily, then one must be
made by means of the cold-chisel and hammer. Look out a

rock whose surface is broken with the holes of the *Pholas*. Taking advantage of these holes as weakening the rock, a piece of the required size can be marked off with the cold-chisel, and then by vigorous chipping can be separated.

If a suitable stone can be found ready to hand in the rock-pool, and it has green weed growing from its surface, you need nothing better, especially if that weed be the thin membrane-like tubes of *Enteromorpha*, for it will continue to grow in the aquarium. But beware of stones with a growth of any of the thick-fronded leathery olive weeds. For a few days they will look well, but then they will begin to decay and melt in slime, with a putrid odour that will assuredly kill everything in a day or two more, and drive you out of the house.

Should you be staying at the seaside only for a few weeks, and desire to see as much as you can of these creatures, yet have no proper aquarium to accommodate them, remember that any vessel not too deep that allows you to look into it will serve your purpose. Even a soup-plate, or an old-fashioned saucer may at times serve better than anything else for observation purposes. But if greater depth be required, a china "slop-basin," or a thin-glass tumbler may be borrowed or otherwise brought into requisition.

To convey Anemones from the sea to a distance, it is best to wrap them lightly in some of the finer seaweeds and put them into a weed-lined box. This is much better than attempting to carry them in water, and will be attended with more satisfactory results.

CHAPTER VII.

SEA-STARS AND SEA-URCHINS.

At low-water, turning over stones and looking into rock-crevices, we are sure to come across members of the *Echinodermata*—the creatures with tough and rough or spiny coverings, popularly known as Star-fish and Sea-urchins. There are many forms of these to be found on the British coasts, though some of them are peculiar to deep-water, and not likely to fall in our way, unless it be their dead bodies washed up to our part of the shore. But we can obtain a fair knowledge of the class to which they belong, from the specimens we can find living their lives in our own littoral zone. Here, hunched up into an almost globular form under this drooping mass of leathery wrack, is the common Five-fingers, Cross-fish, or Star-fish (*Uraster rubens*). Turning him on his back we see the reason for the contracted condition of his five rays: in the hollow thus formed he holds no less than three specimens of the Purple or Dog-winkle. Why? He is a glutton, and is eating those three poor mollusks at one sitting.

Not many years ago we all believed literally the tales that were told of the Star-fish swallowing oysters as large or larger than itself. It was well known that they caused havoc to oyster and mussel beds, and that seemed the most likely way in which the valuable bivalve would be destroyed. Some went so far as to assert that Five-fingers waited his opportunity to catch the oyster gaping, and then slipped in one of his fingers, and so prevented the shell closing. It was left to the imagination to picture that same finger hooking out the native and swallowing it in the approved fashion—off the shell.

The Uraster's mouth is small, and the integuments tough and not capable of great distention; but its stomach is a most accommodating organ, though a very delicate one, and when the Star has come upon food too large to pass through the mouth to the stomach, the stomach passes through the mouth to the food. It surrounds the victim with its fine membrane, pours out its gastric juice, and having reduced it to a fluid condition, re-absorbs the whole, and returns to its natural position inside the Star. That is a wonderful process, but it is quite a common one, and you will certainly catch the animal in the act before you have long been shore-hunting.

This is probably the way in which the securely boxed-up oyster falls a victim to Five-fingers. The oyster's powerful adductor muscles keep the valves closed, and appear to defy any burglariously-disposed creature of its own size; but Five-fingers' gastric juice is a penetrating solvent which paralyzes the muscles and kills the oyster. The elastic hinge then opens the shell automatically, and allows Five-fingers to make an unresisted entrance, and a short end of the oyster.

As we have shaken off the dog-winkles, the Star-fish takes in his stomach for safety, and we are enabled to have a look at his exterior. When we say that he has five rays proceeding from a common centre, we have said well nigh all that is to be said about his form. But the minutiæ of the organs disposed over those rays, and within them—for their interiors form part of the general body-cavity—requires much describing and explaining. The creature has no legs, yet he moves with considerable celerity in any one direction as easily as another, and inequalities of surface present no difficulties to him. And yet the five rays, from their stiffness, are practically useless for this purpose; but on the under surface of these rays are hundreds of pliable and active little suckers, worked by hydraulic power, and it is all one to them whether they have to walk on rock, weed, or glass, up or down, across the floor, or under the ceiling.

Looking at the underside of this Star, we find that each of the rays is deeply channelled along its centre. Only the true Stars have got this channel; the Sand-stars and Brittle-stars have not, neither have they got the wonderful suckers; but along each side of the channel, under each of Five-fingers' arms, there are two rows of soft filaments that bend and wave in any direction, and that end each in a little knob containing a tiny limy plate. By means of this little plate each knob is converted into a sucker, similar to those by which trades-people suspend their goods from the surface of their plate-glass shop-fronts, but worked by water instead of air. There are hundreds of these to each ray, and all act in unison, so that real progress is made when Five-fingers' olfactory sense informs the sucker feet of the direction in which food may be sought.

Ah, you say, has it a nose? No, it has not; but experiments have shown that the entire underside is sensitive to odours. At the tip of each ray there is a spot that is ordinarily spoken of as its eye, but it has no true eye, though these spots are sensitive to light. Its mouth is in the centre of its under-surface, and opens directly into the stomach, which has branches running into each of the rays. The vent for the undigested particles of food and for waste, is on the upper surface.

Near the junction of two of the rays on the upper surface will be seen a round stony knob, which is sometimes taken for the creature's eye. This is not a very wild shot at its purpose, though it is entirely a wrong one, for as placed it certainly does suggest some such function. Its real office could not suggest itself to any person unacquainted with the internal economy of the Star-fish. Looked at through a lens, it will be found to have a number of minute pores in its surface. Strange as it may seem that the Star-fish should require such a convenience, this is really a filter. Scientific men honour it with the important-sounding name of the "madreporiform plate," because its tubes resemble somewhat those of the madrepore coral. I have already referred to the hydraulic system by which the sucker-feet are distended and worked, and this is the "intake" of the supply, as a water-company would call it. Within these is a tube, running near to the creature's mouth on its lower surface, and connecting with a ring of tube that surrounds the mouth, and sends out a branch to each of the five rays. To this branch-pipe all the sucker-feet in a particular ray are connected, and the pressure can be so regulated as to alternately distend the sucker-feet, or to leave them partly empty and flaccid.

Upon one occasion, when I was describing these arrangements of the Stars to a jocular friend, he said the idea of having a big mouth that let in water freely, and a number of minute mouths that let it in slowly, reminded him of the poet Cowper's whim in making a large aperture for his big hares to pass through, and a small one for the little hares. He thought the mouth would have served both purposes; but as I pointed out to him, the water that the Star takes in involuntarily with its food goes into the stomach, where the food is retained and the water strained off by the mouth again. This water would contain grains of sand, vegetable *débris*, and other impurities, which would clog the delicate tubes and spoil a beautiful piece of mechanism. The water that percolates through the minute pores of the stony plate must be pure, and free from all extraneous matter, so that the special supply-pipe is a necessity. The scientific appellation of the sucker-feet is *pedicels* or *ambulacral feet*.

We must not omit to mention organs of another sort that occur in plenty among the sucker-feet, and for many years presented a puzzle to naturalists, who long regarded them as parasites—something foreign to the Star-fish. They are now understood to be pedicels that have been specialised to adapt them for particular functions. They consist of slender flexible stalks, ending in an enlarged head of three claws which normally converge to a point, but they are for ever opening and shutting and taking hold of something. They are analogous to those curious bird's-head organs on some of the zoophytes, to which attention has already been directed. Their function is to take hold of seaweeds and other substances, until the suckers can be got to work; also to keep the sucker-feet clean by removing all matter tending to clog them and impede their efficient working.

The upper side of a common Star-fish is covered with rough or spiny plates, and bosses of carbonate of lime secreted by the creatures, and these take definite patterns in different species.

The Common Star-fish (*Uraster rubens*) is well-known for its rough orange-coloured exterior, and its profusion, in some seasons, upon certain parts of the coast. It swarms on oyster and mussel-beds, and causes considerable annoyance to fishermen, who find it taking possession of the bait on their lines, and so keeping off the fishers' rightful prey.

There is a less common but prettier species, the Spiny Star (*Uraster glacialis*), you may find among the rocks at low-water. It is much larger than the common Cross-fish, and in proportion the rays are longer, their sides more parallel, the upper side more distinctly spiny, and the colour a glaucous green, with variations towards violet. It is more angular-looking than the common species.

Another species is the Eyed Cribella (*Cribella oculata*), which has an upper surface quite free from spines or roughnesses, and of a purple colour.

These Stars go through a remarkable metamorphosis. In the year 1835, Sars, the celebrated naturalist, discovered a peculiar creature about an inch in length, to which he gave the name, *Bipinnaria asterigera*, and classed it among the Jelly-fishes. Nine years later, however, some further observations caused him to reconsider this view, and to regard the creature as more probably the larva of a Star-fish; and in the course of a few years this opinion was confirmed by the researches of Messrs. Koren and Danielssen.

The Sun Star (*Solaster papposa*) is really a glorious creature, with a broad central cushion of rich crimson, from
which radiate from twelve to fifteen arms of the same colour,
but with a band of lighter tint at their base. The upper surface
is covered with a network of slightly raised lines, upon which

SUN STAR.

are threaded, as it were, a great number of little cushions,
supporting erect brushes of spines. It may be found at low-
water, but is more frequently obtained from trammels set in
deeper water, and from the fishermen's lines. It is sometimes
nearly a foot across from tip to tip of opposite rays.

In the bottom right-hand corner of the plate on page 93, will be seen a figure of the Gibbous Starlet (*Asterina gibbosa*), in which it will be seen that the figure of the Common Star has been considerably modified by the partial filling up of the angles between the rays, so that the body appears to be more extensive than the rays. This pretty species—it is represented natural size—is fairly plentiful in rock-pools where there is sand and a vigorous growth of coralline and fine weeds. In such pools it is not easily seen, owing to the manner in which it harmonises with its surroundings. It is covered with a short "pile" of spines, of a greenish-grey tint, with an indefinite shade of brown. It is cushion-shaped; and the underside is channelled from the five points to the central mouth. These channels are bordered with a row of spines on each side, to protect the double range of sucker-feet within.

In the same pools, among the rubbish at the bottom, under stones at low-water, and climbing about corallines and weeds, we shall be sure to find in plenty a little Brittle Star (*Ophiocoma neglecta*), of very attenuated proportions, and not exceeding an inch across, if you can get it to keep still whilst you measure it. It is exceedingly active, and all its tiny rays bend and wriggle at the same time.

The Brittle-stars pass to the other extreme from the Starlet, in modifying the five-rayed plan of the Common Star. Here the creature runs almost entirely into the five writhing arms, which leave but little material for the circular trunk, which looks, in truth, as though five active worms had simultaneously seized a minute button by its edge.

There are several other species of Brittle-star to be found between tide-marks, but they all share, more or less, the peculiarity which gives them the popular name. They are so "touchy" that you need scarcely do more than look at them to cause them to voluntarily snap off a part or whole of a ray, or several rays—and they commonly throw off the lot, if they commence self-mutilation. The amputated members are replaced by new growths, if the Star lives; for sometimes this

PURPLE-TIPPED URCHIN.

FEATHER-STAR.

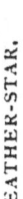

STARLET.

act of renunciation of limbs that offend, is but a prelude to
the extinction of vitality in the trunk.

In the illustration here given of the Granulate Brittle-star,
it will be seen that the rays do not merge imperceptibly into
the trunk, but are attached to it by a kind of dovetail joint on

GRANULATE BRITTLE-STAR.

the upper side.   Below, the arms, at their termination, form a
ring, within which is the mouth, whilst the trunk acts as a
roof above the mouth, and overhanging all round.   The rays
are composed of a series of joints, which allow free lateral
action, or wriggling, but not much vertically.   Each of these

G

joints consists of four little plates, one each above and below, and one on each side. The side plates bear each from five to ten stiff and granular spines of varying length; and short tentacles come out beside the lower plates. These tentacles are not sucker-feet, like those of Five-fingers, but rigid, hooked processes; and there are no *pedicellariæ* with their snapping jaws. The mouth is a very extensive opening, but its area is largely occupied by the five jaws, the free ends of which extend upwards far into the body cavity, and are covered with rows of long, close-set teeth. These teeth, on the five jaws being brought together, must form a wonderfully efficient masticatory apparatus.

One of the commoner forms of these Brittle-stars is the Granulate Brittle-star (*Ophiocoma granulata*), represented in part in our illustration. I have seen crab-pots brought in with this species thickly coating the bottoms inside, and attached to well-nigh every bar; there must have been thousands in each "pot."

An allied species that is more plentiful as an inhabitant of the littoral zone, is the Red Brittle-star (*Ophiothrix rosula*), which will be found sprawling over the under-surfaces of big stones at low-water, in company with the Broad-claw Crab. Of this species Edward Forbes truly remarks :—

" Of all our native Brittle-stars, this is the most common and the most variable. It is also one of the handsomest, presenting every variety of variegation, and the most splendid displays of vivid hues, arranged in beautiful patterns. Not often do we find two specimens coloured alike. It varies also in the length of the ray-spines, the spinuousness of the disk, and the relative proportions of rays and disk ; and in some places it grows to a much greater size than in others. It is the most brittle of all Brittle-stars, separating itself into pieces with wonderful quickness and ease. Touch it, and it flings away an arm ; hold it, and in a moment not an arm remains attached to the body."

Another species, the Long-armed Brittle-star (*Ophiocoma brachiata*), has the rays about twenty times the diameter of the disk, each consisting of three or four hundred joints; so that if one reckons up the four plates that go to make one joint, then adds to these the eight to ten spines on each joint, and multiplies the first total by say three hundred and fifty (the number of joints), and this second total by five (the number of rays), one gets a grand total of seventy thousand pieces, constituting merely the external covering of the rays of this small creature—leaving entirely out of the reckoning the internal bony framework upon which these are placed.

These Brittle-stars go through a peculiar stage of existence, prior to their assumption of rays. When summer is verging upon autumn, their minute larval forms may be gathered in a fine muslin net, from the surface of the sea. Gosse has given a description of this stage with admirable brevity. He says:—

"A painter's long easel affords the only object with which to compare the little creature; for it consists of four long, slender, calcareous rods, arranged two in front and two behind, with connecting pieces going across in a peculiar manner, and meeting at the top in a slender head. On this shelly, fragile, and most delicate framework, as on a skeleton, are placed the soft parts of the animal, a clear gelatinous flesh, forming a sort of semi-oval tunic around it, from the summit to the middle; but thence downward the rods, individually, are merely encased in the flesh, without mutual connection. The interior of the body displays a large cavity, into which a sort of mouth ever and anon admits a gulp of water. Delicate cilia cover the whole integument, and are particularly large and strong on the flesh of the projecting rods.

"The appearance of this most singular animal is very beautiful; its colour pellucid-white, except the summit of the apical knob, and the extremities of the greater rods, which are of a lovely rose-colour. It swims in an upright position, with a calm and deliberate progression. The specimens which I have seen were not more than one-fortieth of an inch in length.

"From this form the Brittle-star is developed, but in a manner unparalleled in any other class of animals. The exterior figure is not gradually changed, but the star is constructed within a particular part of the body of the larva, 'like a picture upon its canvas, or a piece of embroidery in its frame, and then takes up into itself the digestive organs of the larva.' The plane of the future Star-fish is not even the plane of the larva, but one quite independent of, and oblique to it. Strange to tell, the young Star does not absorb into itself the body of the larva, which has acted as a nidus for it, but throws it off as so much useless lumber—flesh, rods, and all!"

Prof. A. Agassiz, however, would have taken exception to that last sentence, for he declared that "the whole larva and all its appendages are gradually drawn into the body, and appropriated."

In the plate on page 93 there are two figures besides the Starlet—the Feather-star and a Sea-urchin. The Feather-star (*Comatula rosacea*) is really a deep-water form, but it has been taken occasionally within the littoral zone, and may occur there in the experience of the reader. It is undoubtedly the most beautiful of the entire group, so far as British waters are concerned, and it possesses a special interest for us, as being the only British representative of the Stone-lilies or Encrinites that so abounded in Palæozoic times that their remains make up whole strata, but of which, until the deep-sea explorations of recent years, no living European species was known. But the Feather-star, as shown in our illustration, had been long known, for in several localities round Britain and Ireland it came up abundantly in the dredge, yet no one suspected it was closely related to the Encrinites.

In the year 1823 Mr. J. Vaughan Thompson, when dredging in the Cove of Cork, brought up a tiny creature less than an inch in length, but which might have been one of these Encrinites, into which life and mobility had been infused. The discovery was hailed with joy by naturalists, and the little stranger was named *Pentacrinus europæus*. Thirteen years later Thompson came to the conclusion that his *Pentacrinus* was only the larval form of *Comatula;* and in 1840 Edward Forbes, Robert Ball, and C. Wyville Thomson were dredging in Dublin Bay, when the dredge brought up specimens of the so-called *Pentacrinus* in a more advanced stage than had been seen hitherto, and behold, some of these underwent the final change in their early history under their eyes: the Feather-star left its stalk and floated off, a true *Comatula*. Sir C. Wyville Thomson has given this interesting account of its progress from the egg condition:—

"The young escapes from the egg a pear-shaped free animalcule, swimming and gyrating rapidly through the water, large end foremost, by four transverse bands of cilia, and by a tail-like tuft of long cilia, which it uses somewhat in the style of a screw propeller. On one side of the body there is a large oval mouth, richly ciliated, and a short curved stomach. After swimming freely in this form for several days, a network of calcareous plates begins to appear, at length making a closed chamber in the wide end of the pear, and extending as a sort of stalk to the narrow end. The stalk now lengthens, and the creature loses its symmetrical form; it attaches itself to a stone or seaweed, and from the free, wide extremity, there springs a little circlet of branches—the arms of the second stage."

On turning again to the illustration (page 93), it will be seen that the Rosy Feather-star, to give it the full title, is possessed of ten rays, or rather five rays each forking into two, and that these branches are *pinnate*, or feathered with little appendages which contain the ova. The ordinary organs are all contained in the central body, and do not extend into the rays as in Five-fingers.

The remaining figure in that plate is the Purple-tipped Sea-urchin (*Echinus miliaris*), which is a well-known inhabitant of rock-pools. It is enclosed in a stone box, which is a miracle of design, for although there is no elasticity about it, and it cannot be stretched, it yet serves the growing Urchin for years, and never cramps him. There is never any necessity for throwing it off, as the crabs and lobsters have to do repeatedly with *their* suits of armour. The nearest parallel to it in nature is the human skull, which although consisting only of a few pieces, enlarges in a similar manner to accommodate the growing brain.

It is remarkable how, in the whole sub-kingdom Echinodermata, all the wonderful variety displayed by the many species is found compatible with rigid loyalty to the dominating "number" principle: in these animals everything is governed by the number five. With a few exceptions, the rays are in fives, or multiples of that number; so are the jaws, the boundaries of the plates, and other details, as may be seen in any of the Stars to which we have alluded.

In the Sea-urchins we get an advance in that direction, for its stone box is built up of nearly six hundred five-sided* plates of lime, securely attached to each other by their edges, and fitting with such beautiful accuracy, that there is not the ghost of a crevice from base to crown of this wonderful cupola.

But if there are no crevices there are many apertures—over five thousand of them in a full-grown *Echinus esculentus*. Forbes, many years ago, calculated there were 3,720 in a *moderate-sized* specimen; and his figures, though used in all the books since his day, do not appear to have been checked. But I have counted the pores in what I should describe as a moderate-sized individual, *i.e.*, one that measured, when denuded of its spines, twelve inches in circumference, and find

---

* Dr. Andrew Wilson, in his "Glimpses of Nature," impresses upon his readers this "pentarchy" in the building of an *Echinus*, but curiously describes the plates as being *six*-sided. He evidently had none but living specimens to refer to when he thus wrote. It is only in a specimen from which the spines and skin have been carefully cleaned that the form of the plates can be seen.

no less than 4,800. The calculation is as follows: ten bands, each consisting of eighty rows of six holes ($10 \times 80 = 800 \times 6 = 4,800$). The specimens that the crabbers take out of their crab-pots, and smash against the rocks, are commonly much larger than this. It always grieves me to see such wonderful structures destroyed in that fashion.

The five thousand pores are in pairs, each pair giving rise to one *pedicel* or sucker-foot, like those we described in the Five-fingers. The ten bands of pores are also arranged in pairs, the bands forming a pair being separated by about five rows of spines, and each pair of bands being separated from the next by twenty (4×5) rows of spines. These intervening spines are borne on two series of long, yet still five-sided, plates; the number of spines to each, in a growing specimen, varying; but from counting many of these I should suppose a fully-grown plate, from the middle of a series, would support twenty spines; at the top and bottom of the series, however, there are only two or three spines to each plate. These spines are not mere rigid outgrowths like the prickles on a chestnut bur; they are beautifully finished pieces of mechanism, with considerable latitude for movement in any direction. Although only about five-eighths of an inch in length, each is a beautiful column in alabaster, tapering slightly to the top, and decorated from near the base with a series of thirty (6×5) parallel rounded ridges. The bottom of this spine is hollowed out and polished perfectly, to enable it to move freely on the polished knob upon which it fits. These knobs are the bosses left on the shell when the spines have been cleaned off; the spines being held to them and moved by a circular band of muscular tissue.

If we look at the underside of the Urchin we shall find the mouth occupying the centre, with five polished white teeth protruding. Although these are not much to look at from outside, they form a large and complicated structure within, which goes by the name of the "Lantern of Aristotle," because the famous Stagyrite appropriately compared its shape to a lantern.

Within we find a set of organs similar to those described in connection with Five-fingers, much of the space being occupied with the water-vascular system by which the enormous number of sucker-feet are worked. The Urchin also possesses a great number of pedicellariæ which keep the upper parts of the huge sphere clean, by passing any particles of dirt from one to the other, until they are passed off altogether. The madrepori-form plate is situate right at the summit of the edifice, near the five eyes and the vent. As its specific name suggests, this urchin is edible; it is boiled like an egg.

The Purple-tipped Urchin (*Echinus miliaris*) is depressed in form, and its outline would represent an oval from which one-fourth had been cut away, whilst *E. esculentus* would repre-sent a circle from which about one-sixth had been abstracted. The skin of *E. esculentus*, when the spines are removed, is red-dish; that of *E. miliaris*, a dusky greenish-grey. *Miliaris* is common in rock-pools and about the rocks at low-water; but *esculentus* is found in deeper water, though, from the frequency with which it is brought in by the crabbers for destruction, rather than throw it overboard where they find it, and from its empty house being rolled in by the waves, it is a fairly common object of the shore.

There is a rarer shore species, called the Purple Urchin (*Strongylocentrus lividus*), which excavates circular holes in the rocks large enough to house itself, spines and all. This is more plentiful in Ireland than on the English coasts; and it is remarkable not only for its excavating propensities, but also because it sheds its thick purple-spines annually, and produces a new crop.

Closely allied to the Sea-stars and Sea-urchins are the Sea-cucumbers, of which we have a number of native species, though many of them belong too exclusively to the deeper waters to be mentioned here. Several of the genus *Cucumaria*, however, may be met among the rocks, at low-water, on our southern coasts. One of these is represented in the

SEA-CUCUMBER.

SIPUNCULUS.

accompanying plate, protruding from a crevice in the rock.   It is the Sea-cucumber (*Cucumaria pentactes*), a species that requires a fair pair of eyes to detect it.   Certainly, when seen for the first time, unless the finder had previously read about Sea-cucumbers, it would never strike him as being a relation of the Sea-stars and Urchins.   There are no spines, no limy plates; instead, the body is soft and molluscous, so that it can progress by its alternate extension and contraction.   But a careful scrutiny of the appendages encircling the mouth might awaken suspicion, for there are ten branching rays, and then it might be noted that the body has five distinct angles, and that these angles are pierced with pores not unlike those of the Urchins, through which protrude sucker-feet.   This, he would consider, constituted a very strong case in favour of their relationship to the Echinoderms; and in this conclusion he would be in agreement with the scientific men, who have, however, also taken the Sea-cucumber's internal arrangement into consideration.   Another point which suggests affinity with the Sea-stars—especially with the Brittle section—is their trick when suffering from want of food or lack of oxygen in the water surrounding them, of throwing off portions of their body, and thus increasing their chances of life by their reduction of the area or bulk that has to be fed or refreshed. The animated Cucumber not only throws off its rays for such reasons, but also its mouth and dental apparatus, and its intestines and ovaries are turned out, and only an empty hollow bag remains.   Should its prospects brighten through the access of food and the oxygenating of its surroundings, it will, in the course of a few months, reproduce these sacrificed organs, and make a fresh start with a new lease of life.   This is a close connection of the tropical Beche de Mer, of which the Malays make Trepang, a very important item in their trade with China, by whom it is used as a choice article of food.

The creature to the left of the Sea-cucumber, on page 103, is the Dotted Siphon-worm (*Sipunculus punctatissima*), formerly included with the Sea-cucumbers, but now relegated by the systematist to the biological lumber-room, whose door is labelled " Vermes," that limbo to which all sorts of creatures are sent who cannot be satisfactorily classified, in the hope that future discoveries may make their affinities more clear. The Siphon-worm has a cylindrical proboscis that is almost as long as its body, and a wreath of simple tentacles around the mouth.

CHAPTER VIII.

SEA-WORMS.

A CHANCE reader picking up this volume by accident, or from curiosity, and opening it at this chapter, will in all probability put it down quickly with the remark, "Worms indeed! and who wishes to read about such disgusting creatures?"

Our prejudices trip us up at every other step we take, and interfere with our seeing and learning much that would interest and edify us. Our notions of worms are suggested by our imperfect knowledge of the common earth-worm (*Lumbricus*) which few persons have properly seen. It is a nasty, slimy, wriggling creature, that spoils the look of the lawn with its unsightly casts, and is a further nuisance in that it disturbs the seedlings in our seed-beds.

Well, as a naturalist I have no great sympathy with this view, for a worm is a wonderful creature; but there are worms and worms, and probably the most sensitive soul who would shrink from a near view of the loathly earth-worm would have his or her interest awakened by a sight of the Rainbow Leaf-worm, the golden-haired Sea-mouse, the cinnabar Cirratulus, or the glowing plumy crown of the Tube-worm. So, too, their imagination may be stirred at the marvellous power of elongation possessed by the *Lineus*, whose full length can only be estimated with difficulty, but which has been ascertained to be something over twenty feet.

There are among them builders—in porcelain, stone, sand, mud—and spinners of submarine webs like those of spiders. Brilliant colours, elegant forms, wonderful structures and mechanism, ease of motion, and symmetry, are among the attributes of the humble sea-worms. Let us look at a few representative forms.

Flat, or nearly flat rocks that are only uncovered by the recess of the spring tides, will often be found to be strangely coated with coarse sand in which are immersed round tubes with their mouths protruding. This is a colony of the common Sabella (*Sabella alveolaria*), which cements the sand together in long tubes, and appears also to spill some of its liquid glue around ; for the spaces between the tubes are filled with sand similarly agglutinated, so that the whole surface of the rock is uniformly coated with sand in which are the sabellæ-tubes. There is nothing to see, so long as the sea remains out ; but when the incoming tide covers this rock it is a sight worth seeing. From every one of these tubes there comes forth a plume of feathers in shape like a funnel. The tubes are fashioned by curiously-modified antennæ, which serve the purpose of a couple of trowels to manipulate the material that has been scooped up by another organ, to mould and smooth it, and make it comfortable for habitation.   •

Its methods of working can be clearly seen by capturing a specimen or two, evicting it from its home, and placing it in a glass vessel with a little clean sand on the bottom. It will immediately proceed to the elaboration of a new tube ; and with that eye for economy of labour and material which characterises the majority of natural builders, it will make the glass serve as the base of its tube, and thus reduce its labour by a third.

The breathing organs (*branchiæ*) of these tube-masons are external, and form a very beautiful object when the worm lies on the threshold of his house and pushes this apparatus out, that his blood may benefit from the abundant oxygen of the ever-moving waters. At first the branchial plume issues very cautiously and with slight pauses and withdrawals ; but finding all safe the *Sabella* at length gets it quite out and expands it to its glorious fulness, delicate in structure, splendid in

TRUMPET SABELLA.   BRUSH SABELLA.   COMMON SABELLA.

colour as the light is variously reflected from the finely-toothed threads.

We must use the lens if we are to get an idea of the structure of this beautiful crown. By its aid we find there are a great number of filaments, each one fringed with finer processes on each side. Fine as these are, they are all hollow, and through them the blood constantly flows, to be brought in contact with, and to absorb the oxygen of the sea-water, which can pass through the microscopic meshes of their walls through which, however, the free cells of the blood cannot pass. In some species these gills are arranged not in circular form but spirally round a central shaft.

Among the numerous species of *Sabella* to be found on our shores, there is one that is not inaptly termed the Silkworm Sabella (*S. bombyx*), and indeed, being a real worm it has more claim to the title than has the caterpillar that is called the Silkworm. This silk-producing Sabella, however, could scarcely be pressed into the service of man, though one could fancy an imaginative writer employing this spinner to make gossamer vestments for sea-fairies, for the material produced is of just the texture a fairy would desire. Not long ago, I introduced to one of my aquarium vases a flat stone that supported a sea-anemone, which I was loth to disturb, and would rather he moved off on his own account. In doing this, one never knows what one may be introducing in addition to the specimen desired, unless one takes the precaution of scraping or scrubbing the stone. A week or two later, I was surprised one morning to find several threads—so clear as to be scarcely visible—running up from this stone at the bottom to a point about four inches up the glass. Next day there was more of it, and so on from day to day the quantity increased, and the older portion became more visible than before, for its extreme transparency passed away, and it became dusty-looking—in fact, cobwebby. By this time it was clear that what had at first looked like purposeless threads and filaments were really part of a quite voluminous tube.

One of these tube-worms, the Trumpet Sabella (*S. tubularia*), is represented in our illustration on page 109. It does not form its tube of foreign material, but of shelly matter secreted by its own body. It does not associate with other individuals of its species as does the common kind (*S. alveolaria*), but attaching the small beginnings of its tube to a shell or stone, it builds by itself and secretes a long tube that gradually rises from its support and attains a more or less erect attitude. From this " coign of vantage " it expands its glorious jewelled coronet, and instantly vanishes far into its depths on the slightest alarm, real or false.

The tubes of another family, the Serpuladæ, resemble those of the Trumpet Sabella in their material, but instead of the semi-erect, free tube of that species, most of the Serpulæ are cemented to shells and stones for the greater part of their length, and are irregularly twisted. There is an important feature, however, which will enable us to distinguish between Serpula and Sabella at a glance. Serpula is furnished with a peculiar organ in shape like a long inverted cone, so placed that it is the last part of the animal to be withdrawn into the tube, which it accurately fits and effectually closes like a stopper. This organ is really one of the tentacles specially developed to serve the purpose of a house-door.

The species represented in the accompanying figure is a very beautiful one, the Scarlet Serpula (*Serpula contortuplicata*). Its scarlet stopper and fine fanlike branchiæ present a splendid contrast with the smooth white, china-like tubes.

Along the sides of these creatures are peculiarly shaped and toothed hooks, and bunches of bristles which serve in lieu of limbs to enable the worm to push out its breathing apparatus and to rapidly withdraw it.

We must look for the Scarlet Serpula on shells and stones that have been washed in from deeper water; but there is a more plentiful species to be found in abundance between tide-marks, sometimes almost completely covering loose flat stones

with its ridge-shaped tube, which earns its scientific name
(*Serpula triquetra*). Of the three flat surfaces implied in
that name one is cemented throughout its length to the shell
or stone it has selected for its freehold. A third species (*S.
vermicularis*) secretes a round tube, but may be readily dis-

SCARLET SERPULA.

tinguished from *S. contortuplicata*, by its possession of a
double stopper with toothed edges.

One of the most plentiful of these tube-making worms is the
Spirorbis, which is to be found everywhere on stones, rocks,
and weeds in the littoral zone. More especially shall we be

struck by its numbers when we observe it thickly studding the fronds of the Toothed Wrack (*Fucus serratus*), for the dark olive hue of the Wrack throws up the dead-white Spirorbis tubes very strongly indeed. These tubes and the animals that form them are very like Serpulæ, but the tube instead of being more or less straight, or merely twisted, is coiled in a flat spiral, like the shell of the fresh-water Trumpet snail (*Planorbis*). Normally these are very flat at the base, and regularly formed, but where (as in specimens before me) they are densely crowded on the Wrack, there is not sufficient room for this regular growth when they get large, and the outer turns of the spiral are twisted aside and greatly distorted. The worm is very like a Serpula, closing its shell with a similar stopper, but the branchial plumes are not nearly so extensive, these rosy appendages being but six in number in Spirorbis.

It is impossible to do much work upon the shore before coming upon some specimens of another species of tube-maker, though of a less artistic character. The probability is that you will turn over a flat stone that is partly imbedded in the sand, and under it will find a furrow with an active worm wriggling through it. On glancing at the stone the explorer finds that he has ruined a habitation by forcibly tearing off the roof which had been cemented to the stone for greater security, and continued for some distance beyond the stone on either side. The tube, as a fact, is of great length, so that the worm, which is not more than six inches long, may have ample room for exercise without going into the dangerous glare of daylight, to be seen by some ravenous fish. This species is commonly known as the Sand-worm (*Arenicola piscatorum*). In some districts it is the "Lug." It is popularly thought to be a favourite bait with fishermen, and it is so described in all the books; but in the part of Cornwall where this book is being written the fishermen do not set great value upon it, though they highly appreciate the Wilfry or Woolfry (*Nereis pelagica*).

The Lug does not produce a very favourable impression when you have turned him out of his burrow, for his very dark greenish hue looks black at a little distance, and his branchial tufts give him a ragged appearance. The fore part of his body is much swollen, but runs off to a point where the proboscis is situated. The branchiæ are attached to about a dozen of the middle rings only, in branching tufts that change from green to crimson. It is a rapid burrower, opening a way through the sand with its proboscis, widening it with the thicker part of its length just beyond, and exuding a mucous cement that agglutinates the grains of sand and leaves the passage open for further use. Its body is cylindrical throughout.

In similar situations we shall find a vertical shaft of sand protruding from the shore, with a kind of halo of fine branching sandy tubes around the mouth. The whole structure will consist either of grains of sand or fragments of shell cemented together on a silky lining. Its mouth is about an inch above the level of the sands, but the tube, if carefully dug out, will be found to extend to a foot or more. This is the home of the Sand Mason or Shell-binder (*Terebella littoralis*); and now that the tide is out the master of the house will probably be lying, like Truth, at the bottom of his well-like structure, and ready to bolt still deeper in the sand if necessary. He is about four inches long, and the most distinguishing feature is a regular mop of pink tentacles around his—I had almost said head, but he has no head, so we will substitute the more correct expression "anterior segment." The gills are much branched, and there is a bright red stripe along the under surface. There are several allied species; one known as the Potter (*T. figulus*) from its choice of mud or clay as the material for its tube.

There is a remarkable worm called *Cirratulus* that lives in stones. Some say he bores the stone, but of that I am very doubtful; but there is no question that he lives in the perforation. Gosse says "under stones," and I have no doubt Gosse

is right; it is sometimes taken under stones, I dare say, for it
leaves its burrow occasionally and sees the outside world.   A
living specimen now before me is in that free condition, having
quitted its stone yesterday, Boxing Day, 1895, and not yet
settled down again.   He is evidently an up-to-date worm, and
goes out on Bank Holidays!   He is about four inches long,
though from his restless wriggling and obvious objection to
assuming a straight form, it is not easy to measure him accurately.   His body proper is of a fine cinnabar colour, and
appears to be hung loosely in a clear outer skin, which is very
roomy in the fore half, sufficiently so to allow the contained
body to curl and twist and double upon itself without affecting
the envelope.   A series of sausage-shaped expansions of this
envelope constantly travel from the rear, forwards, and are
caused by water that has passed through the creature's gills
and is now making its way out along the outer envelope.
Cirratulus has a head, a rather poor one, and a mouth, but it
is not easy to find either, for the segments near the head produce an enormous mop of tentacular processes, many of them
five inches in length, which completely hides the head and
mouth.   These are of the same bright red as the body, and
when they are extended in all directions, and the creature in
a good light is shown to those ignorant of Annelid-beauty as
a worm, it causes a considerable shock to their notions of
worm-repulsiveness.   This shock is not abated when the light
plays on the bristles and a ripple of silvery flashes runs along
them.   In the dark a gentle touch will cause the entire creature to flash with a bluish electric light, which runs also along
every one of the hundreds of finely attenuated filaments from
the head-region.

There is a group of these lowly creatures that are really
magnificent.   They build no tubes, neither do they sink definite tunnels, but they shun the light and lurk under stones, in
the chinks of rocks, and round about the roots of seaweeds.
Such are the Leaf-worms (*Nereis*), of which several are of

great length. They have more or less linear bodies, of equal thickness for the greater part of their length, and consisting of a great number of joints. The head is conical, and adorned with several antennæ. They are carnivorous creatures, and have the proboscis armed with a pair of jaws well toothed. The well-developed feet protrude from the side, and bear gill-warts at their tips, and jointed bristles. One of the most plentiful and striking of these is the Wilfry (*Nereis pelagica*), previously alluded to, a killing bait for sea-fishing, for no fish can resist its glowing play of iridescence. The colour is a pale fleshy-fawn, but with a succession of metallic gleams shooting over it. It is six or eight inches long, and exceedingly active in its movements. Its favourite habitat is the fœtid black muddy sand, rich in organic matter, that collects in hollows between the rocks, or in the mud of brackish creeks.

If you desire a real good day's fishing, spend half of the day before in grubbing for this worm, with bare legs in the rich mud of such a creek; a better plan is to pay somebody a few pence per dozen to get them for you, and save yourself much discomfort.

Another species is the Pearly Nereis (*Nephthys margaritacea*), similar to the Wilfry, but much smaller and running off to a very slender point behind. The warm fawn colour of the upper surface exhibits lively silvery iridescence, very suggestive of mother o' pearl. The large proboscis is cleft in two and adorned with a fringe of greenish processes. The large feet carry each a leaf-like expansion in front of each branch, and tufts of bristles. It occurs chiefly in the sand near low-water.

The Rainbow Leaf-worm (*Phyllodoce lamelligera*) is one of the most glorious of this group of worms, for each of its three or four hundred segments bears a couple of expansive leaf-like plates, which are the breathing organs. These are of a vivid green colour, and on the back of the body proper this hue

changes to blue-green shot with purple and olive gleams.
Its head is rounded, and is distinguished by the tentacles
about it.   This species attains a length of over twenty inches,
but there is, among several others, a small intensely-green
form (*Phyllodoce viridis*) about two inches long, to be found
among the roots of weeds on low rocks.   As this is very
slender and of thin texture, it can be well examined under the
one-inch power of the microscope, when the rowing action of
the gill-leaves, and the extrusion and withdrawal of the
bundles of crystal bristles will be seen.

Another family of these tubeless worms is represented in
the Sanguine Eunice (*Eunice sanguinea*), of which specimens
may be found a couple of feet in length, and of considerable
thickness.   It is green in colour, but
the gill-plates are of a glowing blood-
red.   One edge of these plates is
cut up after the fashion of a comb;
and its head is ornamented by fine
antennæ.   M. Quatrefages has left
a graphic description of this worm
under the microscope, and as that
account has not been greatly hackneyed, I reproduce part of it here.
He says: " We have just placed upon
the stage [of the microscope] a little
trough filled with sea-water, in which
an Eunice is disporting itself.   See
how indignant it is at its captivity;
how its numerous rings contract,
elongate, twist into a spiral coil, and
at every movement emit flashes of
splendour in which all the tints of

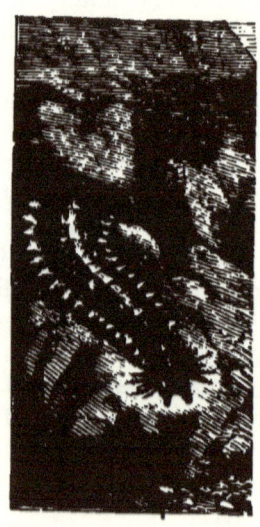

PEARLY NEREIS.

the prism are blended in the brightest metallic reflections.
It is impossible, in the midst of this tumultuous agitation, to
distinguish anything definitely.   But it is more quiet now.

Lose no time in examining it. See how it crawls along the bottom of the vessel, with its thousand feet moving rapidly forwards. See what beautiful plumes adorn the sides of the body; these are the branchiæ, or organs of respiration, which become vermilion as they are swelled by the blood, the course

RAINBOW LEAF-WORM.

of which you may trace all along the back. Look at that head enamelled with the brightest colours; here are the few tentacles, delicate organs of touch, and here, in the midst of them is the mouth, which, at first sight, seems merely like an irregularly puckered slit. But watch it for a few moments; see how it opens and protrudes a large proboscis, furnished with three pairs of jaws, and possessing a diameter which equals that of the body within which it is enclosed, as in a living sheath.

"Well; is it not wonderful? Is there any animal that can surpass it in decoration? The corselet of the brightest beetle, the sparkling throat of the humming-bird, would all look pale when compared with the play of light over the rings of its body, glowing in its golden threads, and sparkling over its amber and coral fringes.

"Now let us take a lens of higher power, and move the lamp in such a manner as to let its rays fall on the reflector of our microscope, and examine a few of the hairs taken from the sides of the Annelid we have been describing. To the outer edge of every foot are appended two bundles of hairs (*setæ*); these are far stiffer than ordinary hairs, and appear to be placed on either side of the animal to defend it from its enemies. A moment's consideration will suffice to confirm this view, for there is perhaps scarcely a weapon invented by the murderous genius of man, whose counterpart could not be found amongst this class of animals. Here are curved blades, whose edges present a prolonged cutting surface, sometimes on the concave edge, as in the yatagan of the Arab, sometimes on the convex border, as in the oriental scimitar. Next we meet with weapons which remind us of the broad-sword of the cuirassier, the sabre, and the bayonet; here are harpoons, fish-hooks, and cutting blades of every form, loosely attached to a sharp handle: these moveable pieces are intended to remain in the body of the enemy, while the handle which supported them becomes a long spike, as sharp as it was before. Here we have straight or curved poniards, cutting-bills, arrows with the barbs turned backwards, but carefully provided with a sheath to protect the fine indentations from being blunted by friction, or broken by any unforeseen accident. Finally, if the enemy should disregard his first wounds, there darts from every foot a shorter but stronger spear, which is brought into play by a special set of muscles, so soon as the combatants are sufficiently near to grapple in close fight."

From the use of the word "feet" in the foregoing it must not be inferred that worms have true jointed feet, like those of crabs or insects, for instance. What are sometimes spoken of as feet in the case of the worm-class are, lateral warts, which carry glassy and elastic bristles in little bundles, like paint brushes; and these are partially withdrawn into a sheath, or pushed out and used like oars with a rowing motion that, all moving rhythmically, send the creature along very speedily, especially when burrowing in sand or mud. To such of my readers as possess a microscope I would advise the careful examination of these bristles, their variety of form, as mentioned by M. Quatrefages, will afford subjects for considerable study; though it is open to doubt whether they are ever used for offensive or defensive purposes.

There is a family of these Sea-worms whose members are mostly characterised by the possession of broad overlapping scales upon their backs, and beneath these are the rudimentary gills, the plates being evidently intended to create currents to supply the blood with oxygen. Two of the most-likely-to-be-met-with of these are *Polynoe squamata* and *P. cirrata*. The former is uniformly pale-brown on the upper side, completely clothed with the large, loose-looking scales, beyond which the three pairs of tentacles, and the lateral organs of touch (*cirri*) project. It is not easy to examine this creature closely whilst it is in a living condition, it is so sensitive to the light, and ever seeking to avoid it. Its chief concern is, "where can I hide?" It possesses four eyes, and its scales are delicately fringed. *Polynoe cirrata* is larger, darker, and its feet protrude further beyond the edge of the scales.

The Sea-mouse (*Aphrodita aculeata*) belongs to this section. It is a species that prefers deeper water, but sometimes comes to shore with a heavy sea. In addition to its scales the arched back is covered with a thick brown felt, thinner in the middle, from which emerge long brown bristles and hairs of yellow and green, that are also iridescent and reflect all the colours of the spectrum.

All the worms we have been considering belong to the Class Annelida, the true worms, with their bodies formed of a long series of rings or segments. There is another group of worms, belonging to the Class Turbellaria, of much lower organization, and generally spoken of as Planarians. Most of them are thin textured creatures that appear capable of almost indefinite expansion, and, on the other hand, they have the power of contraction to a mere speck of jelly. Their voracity is in inverse proportion to their size; and the mouth is situated on the middle line of the under surface, usually not far from the centre of the body, and opening directly into the stomach. The whole of the body is covered with very fine cilia, by whose movements they appear to owe their power of gliding and swimming.   Some of them have a pair of tentacles, though in some instances these are little more than backward folds of the body, and on them or in their neighbourhood is frequently gathered a cluster, or two clusters, of eye-like sensitive spots which, however, do not appear to be very perfectly fitted for visual purposes.   In some cases the mouth is the only opening to the organism, and has to serve several purposes.

The student who would collect and study the Planarians must be gifted with patience and keen sight. The fronds, stems, and roots of seaweeds are suitable places to examine ; also the narrowest cracks and fissures of slaty rocks, where there appears to be no room even for a fine piece of tissue paper. Where such rocks show a loosening of the laminæ, break a portion off by inserting the putty-knife and separate the flakes.   You will see some delicate specimens of the con- ventional worm-shape, but very thin ; you will see mere specks of almost transparent jelly.   Lift these off the stone carefully. How ? Ah, that is a difficult matter, for they are so soft that our clumsy fingers could do nothing with them; but you must be prepared for this, and bring with you a clean camel's hair brush. With this you can pick them up, and by dipping it into a jar of sea-water, and giving a quick rotatory motion to the brush, the Planarian will be dislodged, and will probably settle on the side conveniently for your examination of it with your lens. Any slimy-looking spot of colour that appears upon a stone or sponge you should attempt to move with your · brush, and in many cases it will prove to be a Planarian· that may afterwards so expand as to surprise you with its beauty.

Sir John Dalyell, many years ago, described how he cut up a specimen of the common Black Planaria (*P. nigra*) of our fresh-water ditches, each portion of which became a complete animal; and upon this slender basis appears to have been founded the statement, which is copied in all the books, that worms of this class partly propagate by spontaneous division, in addition to their interesting egg-laying.

Dr. Collingwood, who has paid considerable attention to this comparatively neglected group, doubts this, and I think with good reason. He has never seen this division take place; and I would humbly add that I have kept large numbers of the fresh-water species for years, but never observed the phenomenon, though I have carefully watched for and really expected it to take place—as such fission undoubtedly occurs in the Sea-anemones.

One of the largest and most conspicuous of our native species is the Banded Flat-worm (*Eurylepta vittata*), which is marked with longitudinal black lines on a whitish ground. It has a pair of tentacles in front of the broad, flat body, which gradually tapers away to a point behind. It has a large mouth opening near the centre of the underside.

A more worm-like group is the Nemertea, which is divided into genera, founded on the number or absence of eye spots. One of these is the Red-faced Blind-worm (*Astemma rufifrons*), about an inch and a half in length, with a roundish body, no eyes, and the mouth near the front end.

The Four-eyed Worm (*Tetrastemma quadrioculatum*) is similar in form, but larger, thicker in the middle, and with four eye-spots arranged in a semicircle parallel with the front margin.

The Many-eyed Red-worm (*Polystemma roseum*) has a distinct snake-like head and neck, with many eye-spots in groups around the margin of the head and towards the neck, and in the latter there are two red spots which appear to be hearts. Just below these is the mouth. Viewed laterally the head is wedge-shaped. It is to be found in rock-crevices, and among the rubbish at the roots of seaweeds on the rocks.

The most marvellous, in certain respects, of all these worms is *the* Long Worm (*Lineus marinus**), so long, indeed, that it is all but impossible to give its measurement. It is extremely soft like the others of its tribe, very narrow and quite linear, that is, slender with parallel sides. You will probably find it —for it is fairly common—beneath some deserted shell, resting for the day, away from the light; and it will no doubt be twisted and tangled and coiled upon itself in such a manner as would lead you to say—if you have no experience of its ways—that it were impossible for it or any other creature to disentangle it without many breakages. How any creature can carry on the ordinary functions of life so tightly coiled and twisted and knotted is a marvel. And yet, hopeless as the task of disentanglement appears, *Lineus* accomplishes it without any of those strainings that the juggler puts on when he has been tied up by the sailor, until the confining rope is all knots. Whilst it is day the *Lineus* has no particular desire to uncoil; he is happier as he is, his enormous length more under control and, like an army that is concentrated in one mass, is less open to the assaults of an enemy. But when the fitting occasion has arrived, and *Lineus* wishes to be elsewhere, he solves your

* Better known by its former name, *Nemertes borlasii.*

BANDED FLAT-WORM.

LONG WORM.

difficulties in a way you can scarcely understand, though you see the whole performance. He simply unravels himself; taking the right end of him, and applying a little pressure, he glides off without any fuss, and you see that there is a flowing motion of the black string; no untying, no contortions. He has uncoiled about a foot of himself and laid hold of a stone, a shell, or a weed that distance away, and to the horror of yourself, who hoped now to be able to measure this animated bootlace, he has commenced twisting himself into an equally hopeless tangle at the other end.

He is so remarkably elastic too! You may look at this living Gordian knot and see about a quarter of an inch of the head end protruding from a tight kink; you may watch the kink and certify that no movement takes place in it; yet the head moves away to a distance of five or six inches, simply by the stretching and consequent attenuation of that free quarter of an inch.

The Rev. Hugh Davis many years ago contributed to the Transactions of the Linnean Society an account of his dealings with this Planarian, and as does not often happen to contributions to that useful but technical work, it became much quoted. It was all sober fact, as became the calling of the author and the character of the eminent society to which he communicated the story; but we were greatly amused not many years ago seeing Davis' account of its length, etc., put forward as a specimen of a "traveller's tale," drawn chiefly from the imagination.

Later, but practially identical accounts have been published by Gosse, Charles Kingsley, and others. Kingsley, if we remember rightly, had to defend himself from the charge of shooting with the long-bow, or "slinging the hatchet," and in doing so he said there was so much that was truly marvellous in Nature that it was unnecessary for an author to invent lies wherewith to startle his readers. Yet the story was too much for a well-known and generally well-informed science lecturer, for on the editor of one of the snippety periodicals printing Kingsley's account with the sensational headline, "A living fish-line," and without acknowledgment of the source from which quoted, Mr. W. Mattieu Williams, F.R.A.S., F.C.S., requoted it in his monthly "Gossip on Current Topics," contributed to "Science Gossip," and headed it "Munchausen Science." He coupled it with what he called "an equally sensational account of the latest method of disposing of the dead, by electroplating the corpse," and concludes, "It is not my wont to be presumptuous, but in this case I do venture to suggest that for such revelations the general title of Popular Science should be exchanged for that which I have given, Munchausen Science." Of course, Mr. Williams was a physicist, rather than a biologist, but Dr. Taylor, the editor, professed to have a knowledge of marine biology, and how he could have let Williams' strictures pass without comment or explanation, is more wonderful than the account of *Lineus*.

Davis gave up the attempt to measure the living *Lineus*, but when it was dead he unravelled it without stretching, and found it to be twenty and two feet long. He adds: "I give it as my firm opinion, that I speak within bounds when I say the animal, when alive, might have been extended to four times the length it presented when dead. It is, therefore, by no means impossible that this most astonishing creature may have been susceptible of being drawn out to the length of twelve fathoms, or, according to the accounts of the fishermen, to thirty yards or fifteen fathoms."

I would only add that from my acquaintance with the living *Lineus*, I see no occasion whatever for taxing the Rev. H. Davis or Canon Kingsley with exaggeration. Neither, I think, will my readers, when they have read the following quotation from Prof. W. C. MacIntosh's "Monograph of British Annelida":—"This is unquestionably the giant of the race, and even now I am not quite satisfied about the limit of its growth, for after a severe storm in the spring of 1864, a specimen was thrown on shore at St. Andrews, which half filled a dissecting jar eight inches wide and five inches deep. Thirty yards were measured without rupture, and yet the mass was not half uncoiled."

CHAPTER IX.

CRABS AND LOBSTERS.

THE professional crab and lobster catcher has to provide himself with "pots" and "hullies" for the taking and storing of his crustaceans for the market, and ultimately the table. As we are concerned more with the unmarketable smaller fry, to which the fisherman almost denies the name of crab, we need no such cumbrous paraphernalia; our handy open basket, with its stock of glass jam-jars, is all we require.

Our occupation to-day consists in turning the large stones at low-water in the "long drang," and lifting the heavy tapestry of olive weeds that covers the rocks. In this occupation we shall encounter several species of the crab class, or the Crustacea, as naturalists term that division of the animal kingdom which includes the crabs, lobsters, shrimps, prawns, and barnacles. The crab and the lobster of the fishmonger's shop are creatures that, as adults at least, are chiefly found in deep water, and therefore do not concern us much. But in seeking for other sorts we shall turn out no end of young specimens of the Great Crab, up to three or four inches across the longest part of his *carapace*, as the upper "shell" of a crab is styled in the precise language of science. As this Great Crab, from its occasional appearance on our tables and its large size, is the best known of the whole tribe, we shall do well to use it for a type of the Crustacea, and write a few words concerning it. Any of these small specimens that we can catch under the stones or in rock-holes will serve our purpose, and having taken the precaution to hold his longest diameter between our thumb and forefinger, so that he may not inflict a painful nip with his pincer claws, we shall be able to examine him at leisure.

The most striking feature of the Great Crab (*Cancer pagurus*) is its heavy pincer-claws (*chelæ*), which in a really large male, or Jack-crab, assume enormous proportions. I measured a specimen that a few months since found its way to the cooking pot at home. Across the back, measuring the "shell" only, it was ten and a quarter inches long by six and three quarters from back to front. I took no account of the walking feet, but the big *chelæ* measured sixteen and three quarter inches from the root to the tip, and their girth at the thickest part of the "hand" was eight inches and a half. One of these large specimens of the Great Crab always reminds me of a well-baked pie, when I look at him tucking his legs beneath his roof. It is not alone the substance of his shell and the brown tint that suggests pastry, but there are those deep lines in the frontal margin, marking off the "quadrate lobes" of the scientific describer, that at once reminds you of the marks the cook impresses upon her paste with a fork. Then, of course, there is the pale undercrust; and the resemblance will be strengthened when you observe the voracious Shore Crab, after dining upon a younger brother, holding the empty carapace to his mouth in his pincer-claw, like a piece of pastry, whilst he nibbles at the edge until it is all gone.

So much for this fanciful notion; now let us to business. This shell or carapace of the crab has no more than the merest superficial resemblance to the shells of oysters or other shell-*fish*, falsely so-called. Its relationship is much closer to the horny integuments of beetles and other insects. These are formed of a substance called *chitin*, and of *chitin* also are all the hard parts of a crab composed, with the addition to it, when in a fluid condition, of calcareous matter, which hardens upon a short exposure to the air or water. Where the limb is to bend the calcareous salts are not-deposited, so we find the joints covered with a membrane of soft chitin alone.

The Crustacea belong to that grand division of the animal kingdom known as the Arthropoda, *i.e.*, animals whose bodies consist of a series of variously-shaped segments, the skeleton being external, and giving more definite form to those rings, which are placed edge to edge, and some of which have limbs attached to them. Taking a bird's eye view of the crabs, and seeing only the continuously solid surface of the carapace, it would be difficult to accept this statement; more especially should we stare hard at the crab's back if we were told that the typical number of such rings or segments in the Crustacea is twenty (some authorities say twenty-one). But if we turn the crab over so that we can get a fair view of his smooth white underside, we begin to think there may be something in this ring theory after all, for the undercrust is not solidly continuous like the upper, but marked off by grooves to indicate the seg. ments. The idea is that in the original progenitor of the race the whole twenty segments were distinct and had independent movement, but that in the process of evolution of the various species it has served their purpose in life to have some of these segments soldered together. And so in the many genera into which the vast army of crustaceans are classified, we find great variations in this respect; also in the various functions which the pair of limbs or otherwise modified appendages that spring from each segment is called upon to play.

Under the carapace of the Great Crab are gathered together no less than fourteen segments, nine belonging to the head and bearing appendages transformed into eyes, antennæ, jaws, etc.; whilst five belong to the trunk and bear the great *chelæ* and the four pairs of walking limbs. The remaining six seg- ments belong to the tail (*pleon*), and in the crabs are folded over under the united head and trunk. Among the different groups of crustacea we shall find the widest variations in the arrange- ments of these parts; even in different genera of crabs is this so, as we shall see before we have left the long drang. "Glancing along the whole line of limbs, as the outgrowths from the segments have some right to be called, twenty pairs in number, we find them successively devoted to seeing, feeling, and otherwise perceiving, feeding, and presumably tasting, grasping and striking, walking and digging, swimming and leaping. But although the order in which they act may thus be generally stated, there is not unfrequently a transfer of function from one part of the line to another. The feelers may be employed to assist in swimming or climbing or clasping. The mouth-organs of one group are the grasping weapons of another. The walking legs of one set are elsewhere adapted for swimming. There are also other functions conjugal or maternal, in which the swimming legs or the walking legs may take part, while the breathing apparatus, simple or complicated, may be connected with the mouth-organs or limbs of the trunk or both, or else with the swimming organs of the tail-part, commonly called the pleon."—(*Stebbing.*[*])

What may be called the personal or life-history of the Great Crab is a scientific romance. Once upon a time there was a grotesque sea monster—as big as the head of a good-sized pin —that resembled in a small way a German soldier's spiked helmet, with a couple of huge eyes in front of it, a long jointed tail behind it, and a few bristles around its edge. This creature naturalists recognised as a distinct species, to which they gave the name *Zoea taurus*. It was first taken from the sea by a Dutch naturalist, one Martin Slabber, in the year 1768, but his account was not published until ten years later, whereupon Bosc created a new genus to receive the little oddity. Then there was another sea creature, not much larger, but having a distant resemblance to a lobster, and for this form Leach founded his genus Megalopa. Now it chanced that an Irish naturalist, Mr. J. Vaughan Thompson, nearly fifty years later, thought he would like to verify Slabber's observations, and he searched for the supposed-rare *Zoea*, and found it in profusion. He watched its progress in life, and lo! he beheld *Zoea* cast

---

[*] History of Recent Crustacea ; International Scientific Series, 1893.

its skin and became at once a *Megalopa*. This was sufficiently startling, when the best authorities had agreed that the Crustacea went through no metamorphóses whatever; but continuing to watch and observe, Megalopa was found at its next moult to assume an undoubted crab-shape, and its progress thereafter revealed what has ever since remained one of the most important facts of crustaceology, that no such *species* as Zoca and *Megalopa* exist, but that these *forms* are mere stages in the development of a crab.

As the crab grows and gets too large for its shell, the difficulty of stretching or otherwise increasing the capacity of such a strong-box arises. It cannot be met as in the case of mollusks, by the simple but sufficient method of increasing the length and breadth of the shell by adding new shelly matter to the edge; because the principal part of the crab's internal machinery is in that part of his shell that has no proper edge. There is no help for it—he must do as man does when his garments get too small to accommodate his growing body and lengthening limbs: he gets a new suit. But a glance at his armour-plated condition would suggest that the most difficult part of the business would be, how to get out of the old suit! It might not be such a hopeless task if his limbs were straight and of equal thickness throughout; but in every case the joints are very much narrower than the rest of the limb. Yet, in spite of this difficulty, by the shrinking of the body and its limbs, and by the dissolution of partnership between the upper and lower crusts, the crab, clad in a kind of parchment suit, comes clean out, and leaves his old clothes intact, even to the coverings of the eyes, the antennæ, and the old jaws and mouth-fittings. When the crab emerges from his old home, he is, strange to say, much bigger than that empty presentment of himself, and you might as well attempt to put back the chick into the eggshell it has just vacated as to squeeze the soft crab into his old husk.

Very probably my reader will be so fortunate in some of

his captures as to take a specimen that is on the eve of casting
his shell. He may see, as I have several times seen, the
whole process, and be rewarded with a beautifully clean
cabinet specimen of the crab's shell, perfect in every part.
It only requires careful rinsing in *fresh* water, and drying on
a blotting pad away from the sun or heat, and is then ready to
label and put away.

Many human creatures when they chance to get a new
"rig-out"—to use a nautical expression—are only too anxious
to appear in public, that the cut and colour and pattern of the
garments may be admired, and the wearer—if of the fair sex—
envied; but our crab's paramount desire is to get into a deep
dark hole in the rock, or under a stone, and hide himself.
It is not modesty or shame that thus impels him to hide the
newness of his coat, but the knowledge that he is a wee bit
soft, and too new to meet his own brother, who would instantly
improve the occasion by eating him. He would not like his
own brother to be guilty of the hideous sins of fratricide and
cannibalism at one gulp, and he feels it his duty as his
brother's keeper to put temptation out of his way by seeking
seclusion, until the new crust has set firm and hard.

Here, in this drang, you may frequently find a soft crab in
a hole, awaiting the hardening process; you may as frequently
find a hardened one, or a lobster. They are, in fact, generally
of a retiring disposition, except when looking for breakfast.
Then they quit their holes and cavernous recesses, and come
out on the open rocky bottom where the crabber has dropped
his row of "pots," each with something high and "gamey"
skewered within. Of such a full bouquet is this bait—delic-
ious to the olfactory apparatus of the crab—that he scents
it from afar, and rapidly makes a one-sided progress to the
string of pots. There, within, are the lumps of delight in the
shape of split wrasse, and the osier bars of the pot are so
conveniently arranged that he can easily ascend to the top,
and more easily descend to the interior through the tubular

opening. The prevailing notion is that these pots are so constructed that it is well-nigh impossible for a crab to get out
again; but this is not so, and the fishermen know they must
go round every morning whilst the crab or lobster is still at
breakfast on the savoury viands they have provided, and haul
their pots before he has thoughts of finding the way out. Improved pots have been invented, from which it is impossible
for a crab or lobster to escape, but the fisherman is extremely
conservative, and sticks religiously to the ways and means of
his father's great-grandfather.

Having taken his captures from the pots and thrown them
into the bottom of his boat, the fisherman rows with them to
a protected area of deep-water near the shore, in which each
of the crabbers keeps his own store-pot or *hully*, and hauling
his own particular hully, puts his new captures in. This he
will continue to do perhaps till the end of the week, or until
the merchant comes round with his boat to buy.

Now, having spent so much time over *Cancer pagurus*, we
must leave him, and pay some brief attention to other members of his family—of small concern in the crabber's eyes, but
of equal interest to the student of nature. Under the overhanging masses of *Fucus* that drape the rocks, in the smaller
holes of those rocks and among the stones on the floor of the
drang, we are bound to meet with innumerable specimens of
two crabs that possess no English name. It is true that if
you ask the boys of the place whom you will find at times
among the rocks (and they are the most reliable of local
informants on such matters), they will tell you, with a flavour
of contempt for the crabs, that these are "devil crabs;" but
later on you will find that this term is not specific but generic,
for they apply it to several species that are worthless in their
eyes. In a similar mood, the adult fisherman will tell you
they (and a number of others) are "Zebedees or devil crabs."
Well, Dr. Leach, who founded the genus in 1813, would probably have called it yellow-crab in the vernacular, for he

dubbed it *Xantho*, in scientific language, from the Greek *Xanthos*, yellow. Many of us are more or less colour-blind, and should therefore be careful to abstain from dogmatism in relation to tints, but I should certainly not describe either of the British species of *Xantho* as being yellow, although some specimens of *X. hydrophilus* are certainly yellow*ish*.

Xantho hydrophilus is rather an odd, clumsy-looking creature, owing to the want of proportion between his trunk and the large pincer-claws. The carapace is peculiarly wrinkled, and the margin on the outside of the eye on each side (*latero-anterior*) is marked by four stout triangular teeth. The four pairs of smaller legs (2 to 5) have a row of fine hairs along the upper edge of each joint; and the fingers of the pincer-claws are brown, the moveable one being also grooved on the upper surface. Colour yellowish-brown with darker markings.

Xantho incisus is very like the last, and some specimens will prove difficult to determine with satisfaction. The description of the carapace and its toothed margin will apply equally to either species; but the distinctive characters of this as compared with the last

ZEBEDEE (XANTHO INCISUS).

are that (1) the fingers of the pincer-claws are *black*, (2) that they are *not* grooved, (3) the second to fifth pairs of legs instead of having the fringe of hairs all along their upper margin have only the third (or longest) joint of each leg so decorated. The second and third points are, I believe, reliable —the first is not. I have seen many specimens with the fingers of a paler brown even than the general hue of the big claws, and have such a specimen alive before me as I write.

In the course of our stone-turning we are likely to come upon a little purplish-brown crab, about an inch across the carapace, and bristling all over with hairs and spines. It is known to the naturalist as *Pilumnus hirtellus*, but none of the writers on crabs appears to have troubled about a popular name for it, so it is incumbent upon me to supply the deficiency. For the purpose of communication with my readers, I therefore dub *Pilumnus hirtellus* with the nickname or alias of Hairy Crab. The front of the carapace is cut up into a number of teeth much sharper than those of *Xantho;* in fact,

HAIRY-CRAB.

in comparison with those, these of the Hairy-crab are spines. One of these spines protects the orbit of the eye, and there are four others on each side between it and the base of the pincer-claws. The pincer-claws have a very robust appearance in comparison with the size of the trunk, being thick and rounded; one is usually larger than its fellow, but it may be either the right or the left. The smaller of the two is covered with tubercles on the upper parts, the larger is smooth. The smaller legs are very hairy indeed, and similar hairs are scattered over the carapace among the short down with which it is covered. It is common all along the Southern and Western coasts of England and around Ireland, under stones at low-water, though by no means so abundant as *Xantho*, and others we have to mention. Bell, in his " History of Stalk-eyed Crustacea," almost seems to question Dr. Leach's statement that it is found at low tide mark, for he adds, " those which I have obtained have been from deep water." Dr. Leach, however, was quite correct in his statement, and Bell could easily have substantiated it, as we have done.

We shall not be long at our work before we meet with far too many examples of the Common Shore Crab, Green Crab or Harbour Crab (*Carcinus mænas*); young specimens of which will scuttle away sideways with marvellous alacrity, but bigger examples will at once put up their heavy hands and challenge us to fight. Everybody that has been to the sea-shore knows this crab, for even if entirely void of curiosity as to the wonders of the shore, *Carcinus mænas* will not be ignored. Whether the shore be sandy or rocky, or of that nondescript character that pertains to many harbours, a mixture of sand, stones, and domestic rubbish, this crab will be seen strolling along at a little distance from the water. All know its mottled greeny-yellowy-brown back, and the strength of its sharp nippers! There is only this one member of the genus, so that there is little danger of confusing it with its nearest relations. It most closely resembles certain of the swimming-crabs (*Portunus*), to be described hereafter, but may be readily separated from them by glancing at the terminal joint of the last pair of feet. In *Portunus* this is flattened out as though it had been beaten on an anvil until it was very broad and very thin, to serve as a swimming plate. In *Carcinus*, though the smaller legs are obviously compressed, this last joint of all is stout and runs off to a rounded point, more suited for obtaining a good hold of a sandy bottom than for swimming. We shall find it frequently under both weeds and stones. It is an omnivorous feeder, accepting fish, flesh, or fowl; stealing bait from the fisherman's lines and from his crab-pots, disfiguring the fish which has been already caught on spillers, and, worse than all, causing great havoc among the young oysters that have been laid down in the beds, by eating them, shell and all. They are said to form an important article of food along the shores of the Adriatic, and they were at one time not unknown in the London markets. Leach says that in his time (early in the century), immense quantities were eaten by the London poor. Whether there is any considerable trade of this kind now I do not know; but I remember how more than thirty years ago I considered them very sweet and tooth-some, and used to go as a boy to buy them, all alive, of an old woman in one of that intricate maze of courts and alleys that then existed where now the Royal Courts of Justice stand. I think they were sold at about eight or ten for a penny. Had they not been sold alive I should probably never have desired to have them.

When throwing aside the heavy bunches of *Fucus* that hang over the rocks, in order that we may see their surfaces, we shall

VELVET FIDDLER.

catch sight of a more pugnacious crab even than *Carcinus*, leaping, rather than running sideways, with such rapidity that we need to be smart to catch it. Aye, and we need to have a little nerve, or the Velvet Fiddler will alarm us into letting him pass into the oblivion of the seaweed jungle, or one of those rock-crevices which always seem to be in the right place to afford sanctuary to a poor hunted crab. Most crabs are so flattened that these cracks seem specially provided for them, whereas the evolutionist will tell you it is the rock-haunting crabs that have become specially adapted to find salvation in these asylums. This is the crab we alluded to especially when speaking of the likeness between the swimming-crabs (*Portunus*) and the Shore-crab. The Velvet Fiddler (*Portunus puber*) is one of the swimming crabs; this may easily be seen on reference to the hindmost pair of legs, as already indicated. The Velvet Fiddler gets the two words of his queer name from two distinct characters. He is clad in a dingy suit of velveteen, which appears to be much the worse for wear—rusty, and in places the nap is worn right off, probably by too much squeezing into tight places in the rocks. On his limbs the velveteen is marked in such definite patterns, that we feel inclined to abandon the hard-wear theory, and to fall back upon one of natural artistic adornment. He is really a very fine fellow; his legs being covered on the upper sides with this velvet pile, with the exception of certain longitudinal raised lines of polished blue-black. The square-looking back of the carapace has a similar smooth raised border, with two raised lines of the same character below it. Then all the smaller legs have the longest joint fringed along the upper edge, but the hindmost pair in addition have a close broad band of stiff feather-like fringe standing out all round the three last joints. The last two of these are flattened out to such an extremity of thinness that there seems to be no room for living flesh within. The pincer-claws are not so heavy or robust as those of the species we have already considered. They are more uniform in thickness, more elegant in their slim tapering, so that the members of this genus are often called Lady-crabs. The upper surface is velvety, picked out here and there with blue, and the hand, with its fixed nipper, is decorated below with white and blue tubercles. The moveable nipper is finely ridged, and both of them have a fine row of teeth. Then these pincer-claws are well-armed with long sharp spines; the *antero-lateral* margins of the carapace are finished off with five sharp curved spines on each side, and the space between the eye-orbits are similarly protected, but with thinner, straight spines. The large round eyes are a pair of gleaming rubies, and the tough skin that hinges the joints of the limbs together is of the same hue as the eyes. Such is the appearance of the living Velvet Fiddler; the museum specimens lack much of his brightness and beauty.

The name of Fiddler has been given to him, according to Mr. Gosse, "because the see-saw motion of the bent and flattened joints of the oar-feet is so much like that of a fiddler's elbow." You will, I am sure, agree that this is a satisfactory explanation when you see the Velvet Fiddler flinging these feet about in a perfectly unnecessary, and in-effectual manner, considering that he is out of water. When we disturb him during our exploration of the drang, he puts up his pincer-claws in similar fashion to the tactics adopted by the Shore-crab; but we are not to be alarmed in that manner. Pretending to hit him between the eyes with one hand, we slip the other behind him, and catch the longest part of his carapace between our finger and thumb, and his kicks and threats are thrown away.

There are seven other species of Swimming-crabs belonging to this genus, *Portunus*, found in British waters, but as they all inhabit deep water, and can be obtained only with the dredge, or by arrangement with the crabbers, who regretfully find them in their pots, they are not likely to thrust themselves on the notice of the shore-naturalist.

Gazing into the rock-pools, an observer who was acquainted with molluscan life, but not with the Crustacea, would be aston-ished at the marvellous rate at which winkles, dog-whelks, tops, and other shells move over the bottom; but if he lifted one of these he would discover that the builder of the house had given up possession, and a tenant had taken it for a term. This tenant is one or other of a dozen species of crabs known indiscriminately to the great British Public as *the* Hermit Crab or Soldier Crab. The fact that it shuts itself up in a solitary cell is sufficient to account for its name of Hermit-crab; and a strong tendency to wage war upon a fellow crab, who may

live in a slightly larger shell, is probably the reason for its
military name. The Hermit-crabs are among the curiosities
of crab life—though for the matter of that, so are all crabs.
If there were but one species, we could say it was singular in
the fact that the carapace is reduced to the smallest proportions, and the greater part of the crab's body without a shell
of its own secretion. Nature has been unkind to it in this
respect, so the first thought or prompting in the baby Hermit
is to look around for a deserted gasteropod-shell. It must be
an exceedingly small one to fit him, but he will find plenty
such. It has been a matter for considerable debate whether
the Hermit is content with an abandoned shell, of which the
builder is dead, or whether he first murders and eats the
original owner, and then takes possession of his victim's real
estate. It is remarkable that naturalists should raise such a
question, for anyone who has had any acquaintance with
mollusks must know that if a Hermit-crab were to kill, say a
Purple, a Top, or a Winkle, he would not be able to get the
dead body cleaned out of the shell until putrescence had
loosened the muscular attachment. The Hermit could not
wait for this process, and therefore I imagine this theory must
stand aside until observers have actually seen the crabs in a
state of nature forcibly ejecting the mollusk, and appropriating
its shell. But it is pretty certain that the Hermits do rob each
other of desirable shells, not always with good judgment. A
Hermit in my possession lived in a large Top-shell, but
coveted a smaller, though large Winkle-shell, which was inhabited by a brother Hermit. For about a week these two
were dodging and chasing each other, but to no purpose, for
each is powerless to make any impression when the other
suddenly shuts himself in his shell with a snap, leaving only
the tips of his claws blocking the entrance. However, by
some means he got his brother ejected, and eaten by a Shanny;
he quitting his own commodious Top-shell and putting on the
Winkle-shell. He was evidently trying hard to persuade himself

K

that it was a splendid fit and most becoming; but the whole
business was absurd.   The shell was so small that it did not
protect his soft parts, and in case of danger he could not defend
himself from an attack in the rear.   To add to his troubles
he cast his natural shell, and was, of course, much larger than
before.   For a day or two he still pretended that he lived in a
sufficiently roomy house; then I suppose the pressure on his
abdomen became awkward at dinner time, for he publicly
owned up that he had committed an error of judgment, quitted
the Winkle shell, and resumed possession of his old top-coat,
though this necessitated another murder.   After he had vacated it a much smaller individual took possession, but as he
fitted very loosely it was no very difficult matter for the previous owner to have him out "by the scruff of his neck,"
and give him his quietus.

The most familiar of the Hermits is *Eupagurus bernhardus*,
the Common Hermit-crab, but we are not likely to find full-
grown individuals, which keep out in deep water.   When full
grown, they are about five inches long, and house themselves
in large Whelk-shells.   The characters by which this species
may be distinguished are: the right pincer-claw (*cheliped*) is
usually much larger than the left, and the plentiful granulations
of its surface are almost large enough to be described as tubercles; the last joints of the second and third pairs of legs are
edged on the upper side with spiny teeth, and they are a wee
bit twisted.

Prideaux's Hermit-crab (*Eupagurus prideaux*), is so-called
because Dr. Leach, who first identified it as a species distinct
from *E. bernhardus*, received it from his friend Prideaux, who
had taken large numbers of it in Plymouth Sound.   The
granulations of the pincer-claws are much smaller than in
bernhardus, and whereas the next joint to the pincers in the
latter species has its inner margin decorated with a row of
spines, those in *prideaux* are innocent tubercles.   Then, again,
the second and third pairs of legs are nearly smooth, and their

THE HERMIT-CRAB AND THE CLOAKLET ANEMONE.

last joints have no twist, but instead have a groove carved in each side; the eye-stalks are stouter and the inner antennæ longer than in the Common species. It does not attain such large proportions as *bernhardus*. An interesting point in the natural history of *prideaux*, is the friendly relations subsisting between it and a peculiar species of anemone—the so-called Cloaklet (*Adamsia palliata*)—which attaches itself to the shell serving as the Hermit's cell, and spreads its base out in two lobes, that almost encircle the mouth of the shell. There is no doubt that this *commensalism*, as such alliances are called by naturalists, is of advantage to both parties to it: the anemone is thus brought into contact with food at the Hermit's own table, so-to-speak, and the crab may be in turn protected from the cavernous jaws of fishes, whose gorge rises at the nauseous odour of all anemones. Several such alliances are known in connection with other species of Hermits.

To return to our overhauling of stones: this should be done with care, especially when we are dealing with large masses. I have, when serving my apprenticeship at this kind of work, years ago, had the misfortune, on more than one occasion, to so miscalculate the weight and shape of a large stone, that it has fallen with greater force and in a different direction from that expected—and my toes have been on the spot where it fell! But apart from such accidents, the stone must be turned sharply, or the queer creatures which Nature has specially contrived for living beneath it, will vanish into holes, under other stones, in the sand or mud, or in some other manner.

Among those that require a sharp eye to see them is the Hairy Porcelain Crab, Shaggy Flat-crab, or Broad-claw (*Porcellana platycheles*). Here is his portrait, but it is only fair to the reader I should explain that, like many other portraits, it was taken after the subject of it had been carefully washed and brushed up. *Platycheles* is a ragamuffin, a crustacean mud-lark. There is none other like him in the whole range of British crab-life, though several are fond of dressing themselves up in a

variety of living rubbish; but they do not get themselves so be-
daubed with mud on a coast where mud has to be searched for
if wanted. He has really made it one of the objects of his life to
collect that mud, particle by particle, and entangle it in the luxuriant crop of hair with which he is covered. He is a little fellow
—only measuring about half an inch from back to front edges

SCALY SQUAT-LOBSTER.   BROAD-CLAW.

of the carapace—and I suppose, were he built upon the same
plan as other crabs, he would be smaller, if only the same
quantity of material were to be allowed; for he is flattened out,
and looks as though he had at one time formed part of a
travelling show and the fat woman had sat upon him. His
body is flat, but his pincer-claws are flatter, and the area of

each of the latter is equal to that of his carapace ; they are enormous. And yet, if he had the sense to keep still when the stone is overturned, you would probably fail to see him ; he sits so tightly, and presses the cleaner side of him to the stone. But he has that fatal *crabbiness*, the desire to fight, and whilst he is sidling off somewhere, he thinks he may as well give you a nip, and he puts up one of his massive-looking pincers, and grips your finger with spirit. With your other hand you grip the offending pincer, and say, "Aha! my friend, you've caught a Tartar this time; let go!". He does, but instead of loosing his hold on your finger, he just touches a spring or some other mechanism, and separates his claw from his body without any compunction whatever, whilst his other claws and his body go sliddering off beneath the stone again.

If you catch your Broad-claw young, you will find that his upper surfaces are of a ruddy-brown tint, with hair to match, but when he has got this well filled up with filth, he might pass for a daub of mud. Hold him over on his back, if you can, and you will understand why he is called Porcelain-crab. He is smooth and comparatively clean beneath, and his under surface is of a creamy-white colour.

Broad-claw has an equally odd-looking relative, the Minute Porcelain Crab (*Porcellana longicornis*), which really belongs to deeper waters than our researches at present extend to, but one or two can usually be found under, or among, the stones at extreme low tide. Its colour is red, and its carapace comes very near to being circular. It has not that depressed appearance that makes you pity *platycheles* for having to support such heavy stones upon his back; in truth, the circularity of the carapace, its convexity, and the fact that it has some depth as well as breadth, makes it appear almost rotund. Its larger pincer-claw is almost three times the length of the carapace ; the other about one-third less, and not nearly so thick. They are both rugged in character, and convex, the larger being slightly keeled on top, and the lesser strongly

keeled and grooved. The antennæ are very long, a circumstance to which the creature owes its second name. There is
very little hair about this species, and consequently he is able
to keep himself clean and neat.

In close alliance with the Porcelain-crabs is a group popularly known (as far as they are known at all, which is but
slightly) as Squat Lobsters. They are not lobsters however,
though the long slender pincers, the elongated carapace, and
the lobster-like tail all contribute to the likeness. The most
plentiful species is that figured on page 148, with Broad-claw,
viz. :—

The Scaly Squat-lobster (*Galathea squamifera*), which we
shall find freely under the stones at very low-water in our
drang. He is a very lively fellow, who objects to too much
publicity, and is very anxious to get into a hole or under
another stone the moment you lift the roof off his former
retreat. He shoots backwards in true lobster fashion, his
pincers held straight out in front. If, however, you interfere
with his retrograde movement, the nippers will not be trailed
uselessly, but raised and brought into action. Like Broad-
claw, he does not set great store by a limb or two, and will
willingly part with several as the price of liberty. In colour,
squamifera is very dark olive, the carapace covered with
waved lines across it, said lines being evenly fringed with
short hairs. Similarly fringed scales occur plentifully over all
the legs. The carapace begins in front, with a distinct beak,
and an awful array of fixed bayonets. The first of these is a
stout sharp spine in the very front, and behind it on either
side just above the eyes is a series of four similar spines slightly
curved, of which the first is the largest, and the fourth very
short. Along each side of the carapace is a closely-set row of
spines, and the outer edge of the " hand " is protected in a like
manner. The next three limbs have smaller spines upon their
upper margin, and of larger size, on what might, from its
apparent position, be popularly regarded as the knee. All

these spines, wherever fixed, agree in having red points. But these particulars are not sufficient by themselves to distinguish this species from certain of its congeners, and I am compelled to ask my readers to enter into some minute, and I fear to them, tedious details of description. Of the various appendages to the segments comprised in the head of these crustaceans, some constitute the eyes, antennæ, and jaws. Outside the jaws, and immediately between the pincer-claws of *squamifera*, is a pair of appendages called the third pair of maxillipeds or footjaws, with long hairy fringes to the extremities. Study these carefully, for from these we can tell at once which of three species we are looking at. Each of these mouth-organs, like the larger legs, is made up of seven joints; but it is not always easy to reckon these up from the base, because sometimes a joint is hidden or coalesces with another. If now we commence at the other end, calling the top-joint No. 7, and reckoning backwards, we shall have less difficulty. To save further description, and to make easy of reference, I have drawn up a table of distinguishing features for the British species of Galathea :—

Specimens having branch (*epipod*) from basal joint of pincer-legs and two next pairs of legs	Third footjaws (*maxillipeds*) with 3rd joint shorter than 4th „ „ longer than 2nd	*squamifera*
	3rd joint equal to 4th. „ „ shorter than 2nd	*nexa*
	3rd joint longer than 4th	*dispersa*
Branch (*epipod*) from pincer legs *only*		*intermedia*
No epipods from either pair of legs	2nd joint longer than 3rd	*strigosa*

The Spinous Squat-lobster (*Galathea strigosa*) has spines on his hands along both the inner and the outer margins; and the antennæ are so long that if extended over the back they will reach for some distance beyond the tail. Its colour is inclined to red, with spots and lines of blue. These are the only two we are *likely* to find in our stone turning, and even *strigosa* appears to be more at home in deeper water. According to Couch and Spence-Bate, *dispersa* is the commonest form in Cornwall below the low-water mark. *Nexa* is also a deep-water species.

At extreme low water (spring tides) one may be so fortunate among these rocks to come across a stray lobster or two. Just outside you can see the corks which mark the ends of the long series of lobster pots that are put down to catch them, so that it is not very far for them to stray up to this level. I think my readers could be trusted to know the Lobster (*Astacus gammarus*) if they saw it, without bothering them with a description? Probably they would not be expecting to see a creature with a coat of the same colour as the uniform of a grenadier guard, instead of blue-black relieved on the under-side by dull orange. They may also be trusted to know the Spiny Lobster, Crawfish or Greek (*Palinurus vulgaris*), with its very horrid carapace of purplish brown, its lack of heavy pincer-legs, its red-tinted white legs, and its long, thick and strong antennæ. If you do not come across either of these at low-water, you may see them when the crabbers bring in their catches. Their boats should be watched as they come in each morning, for you can frequently pick up deep-water specimens of *Echini*, spider-crabs, and so forth, that have dropped out of the crab-pots into the boat.

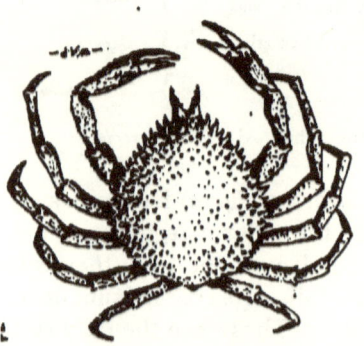

PRICKLY SPIDER-CRAB.

On our south-western shores you will see, brought in by the crabbers, or occasionally at liberty among the rocks, a rough, long-legged fellow called the Prickly Spider-crab, Corwich, or Gabrick (*Maia squinado*), with a convex carapace of oval form, the broadest part behind. His pincer-legs are but little

THE MASKED-CRAB (MALE).

thicker, though much longer, than the others.   On that account he is not greatly esteemed as merchandise, but his flesh is far sweeter than that of the Great Crab.   He is a creature of slow and languid habit, who takes as much pains with the "get up" of his carapace as a lady does with her hair or her bonnet.   His notion is to make it look like a rough piece of rock, with its characteristic flora and fauna, and to this end he takes cuttings of plants, sponges, ascidians, and anemones, and giving them a lick with his lips, as though they were postage stamps, he carefully sticks them in the valleys between the spines and tubercles on his back, adjusting them by means of his conveniently long arms.   The seeker after zoophytic treasures might look in many a worse place for them than on the Gabrick's back.

We have now done as much as possible with the crabs of the rocky shore, and must shift our ground for a while to the flat sands that run out from the upper part of the bay, and taking advantage of the very lowest tides, must go, armed with trowel or spade, to dig in the treacherous sands.   Many things we may find other than those we came specially to seek, and those we specially want just now may not come to light; still it is in the sand we shall find the Masked Crab and the Angular Crab, if they occur in the district.

The Masked-crab (*Corystes cassivelaunus*) has a carapace that is much longer than it is broad, almost elliptical in out-line, and so marked with depressions that some specimens present a remarkable likeness to a human face, more especially so if the crab is held in a way that will accentuate the promi-nences by casting small shadows.   It is prettily coloured with yellow and red.   The male has deeper tints than the female, and his pincer-legs are much longer than hers.   Their habit is to burrow into the sand in rather deep water, and lie buried, with only the tips of their long antennæ at the surface.   These antennæ are furnished with a double row of hairs throughout their length, and by placing the antennæ so close together that these hairs interlock, a tube is formed through which the crab can draw in the current of water necessary for respiration. After storms, great numbers of this crab are sometimes cast up on the shore, dead.

Another crab of singular aspect is the Angular-crab (*Gonoplax rhomboides*), so-called on account of the many sharp angles of the flesh-tinted carapace. Its pincer-legs look as though they had been drawn out when the animal was soft, for in the adult male they are quite four times the length of the carapace—in the female and young male they are much less. Another distinction of the sexes will be found in the colour of the moveable finger of the pincers, which is black in the male only. The eyes are mounted on such long stalks that they reach nearly to the sides of the carapace, which run out into a long sharp spine at each front corner for the protection of the eyes. These are mounted very much like the eyes of the Racer-crab (*Ocypoda*) of other lands, and they are used for a similar purpose. The footstalks are erected so that the crab can see over a wider extent of territory, and behind as well as before. They appear to live in excavations in the mud on our southern and western coasts. They are much esteemed as food by various kinds of fish, and many specimens have been taken from the stomach of the cod particularly.

If it be desired to keep living crabs for the purpose of observing them, a shallow vessel will be found the best; or at least, a vessel in which they can easily get into shallow water. Provision should always be made whereby a crab can climb right out of the water, yet so that he cannot get out of the vessel; otherwise he will wander all over the house, and either get stepped upon, or get dried up in some obscure corner. It must be remembered that the crab consumes much oxygen, and if specimens of any size are put into tanks containing more delicate creatures, much harm may result. It should also be borne in mind that they are of ravenous and omnivorous appetite, and your choice specimens of soft-bodied

ANGULAR-CRAB.

NUT-CRAB.

creatures will not be held sacred by the crabs. We should therefore advise a separate receptacle for crustaceans; and some of the smaller, more delicate kinds, must be kept each in their own vessels. The smaller species will probably be able to pick up sufficient food from the minute animal and vegetable life of your tanks, but the large ones will require to have food specially provided for them. Small pieces of fish will be found the most convenient for this purpose, and it will be more highly appreciated if it be not too fresh. Like the slum-boy who could not relish farmhouse eggs because they were deficient in flavour, the crab prefers his food to be kept for a time.

CHAPTER X.

SHRIMPS AND PRAWNS.

POPULARLY there are about three British species of shrimps, including the Prawn; and the reader whose knowledge of our Crustacea is slight will look for a very brief chapter this time. But he who has paid a little attention to this group will know that we have a difficulty before us in giving anything like a reasonable account of British shrimps without the chapter running into a book. However, our task is greatly lightened for us by the fact that many of these are to be found only in deeper water than lies within the littoral zone, and therefore must be excluded from our survey. By a course of proceeding then from the known to the unknown, we would call attention to the largest of the well-known trio.

The Great Prawn (*Leander serratus*), which we fear is best known in its brilliant red colouring, as seen on the breakfast tables of the well-to-do, and in the shop of the first-class fishmonger. Neither of these places offers great advantages for the pursuit of natural history studies so far as the external appearance of living creatures is concerned. It is in the rock-pools that we must make our acquaintance with the noble Prawn in all the glow and glory of life and activity. It is true he then lacks the fine colour of the boiled article, but he has the greater beauty with which Nature has endowed him; and when you have seen him in his native haunt you will confess that we have not misused the term "noble" in applying it to the bearing of the Prawn.

Not many years ago a learned Professor wrote a book on the sea-shore, and. in it stated, among many other curious things, curiously said, that the Prawn could scarcely be called a shore animal except in its younger condition. Probably he had got most of his natural history from the University, and his Professorial dignity would not allow him to go on hands and knees beside a rock-pool that he might learn of the living creatures there: for in all the pools on a rocky coast, Prawns of all sizes, including the giants of the species, are very plentiful.

The young prawns, though somewhat lacking the grandeur of the older ones, are more beautiful; their shelly armour is so crystalline, and their flesh is so clear. But with adult-growth comes a thickening of the armour-plates, and a very pale brown coloration produced by the greater density of the muscular tissue with which it is principally filled. You cannot help being struck with the pretty colouring of those limbs which the late Thomas Bell called the Prawn's hands; these are the limbs that are furnished with nippers or pincers at their ends, of which the Prawn has two pairs. The first two are very delicate organs, and are only used for delicate work. The second pair the Prawn always carries in front of him, ready for action, but the first are carefully folded up and held close under the jaws. In an aquarium where you have introduced a mossy-looking stone from low-water, you will see the Prawn ranging over it and picking up with his smaller pincers some minute objects that his eyes enable him to see, but which we cannot make out without a lens. In the securing of larger masses of food the heavy "hands" would be employed, but to convey small particles of the mass to the mouth the smaller hands are brought into requisition, and very daintily they perform their work.

The Prawn resembles the crab in his bold, fearless spirit, and it is of little moment to him how he comes by his food. In the rock-pools, as in the aquarium, I have seen him pull some tit-bit out from the depths of an anemone's mouth without the slightest ceremony. He does not quarrel—not a bit. He just walks up to the anemone, and keeping his body clear of her tentacles by means of his spindly walking legs, reaches to her mouth with his widely extended pincers—the larger pair. There is such a matter-of-fact, business-like air about his action that you would scarcely be surprised if you heard him say, " Hallo! what have we got for dinner to-day? Young goby, eh! Thanks; I'll take a little!" and you see the luckless goby that has been stung by the anemone quickly withdrawn from her throat and taken off to the Prawn's den beneath the big stone at the other end of the pool. In all probability, if he has happened to arrive just too late, when the anemone's meal has entirely disappeared from sight, you will see him giving a sly tweak to her tentacles.

The young ones swarm in the pools, and you have only to disturb the drapery of weeds that lines the wall to see a number of them come out into the middle; but the big fellows, of four inches and more in length, keep close, as a rule, in a hole or under a stone. Oftentimes a huge stone that cannot be lifted will be found in a pool supported upon other stones, or kept away from the floor by the concavity of the bottom. A thin stick introduced beneath that stone and moved from side to side will cause several splendid specimens to emerge from obscurity. It now remains for the disturber of their peace to show his activity by catching them: by no means an easy thing to do.

I have already dealt in the previous chapter with the principle of construction in the Crustacea, and the intelligent reader can easily apply the description there to the case of the shrimps here. As in the crabs and lobsters the eyes, antennæ, and various parts of the mouth are all modified feet.

I would strongly advise my readers to catch a full-grown Prawn, kill it by immersion in fresh cold water, cut the body through at the junction between the helmet-like carapace and the first plate of the abdomen, when the entire contents of both head and body can be cleaned out, and the space filled with white cotton-wool. When thoroughly dry the two parts

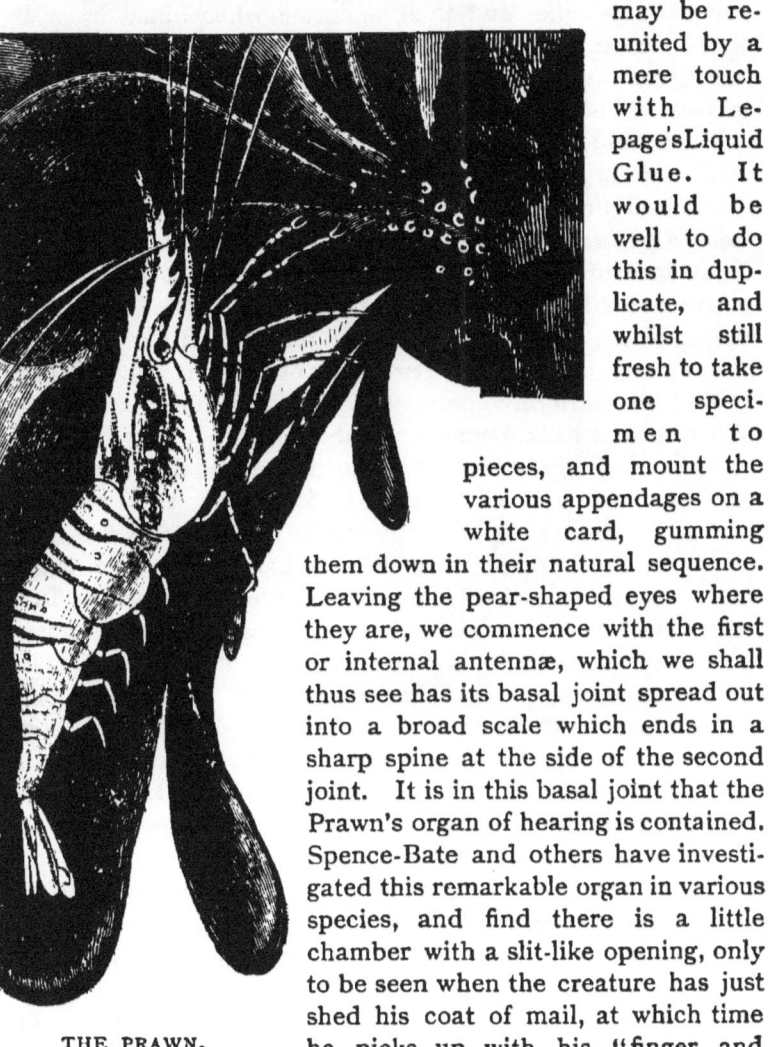

may be re-united by a mere touch with Le-page'sLiquid Glue. It would be well to do this in dup-licate, and whilst still fresh to take one speci-men to pieces, and mount the various appendages on a white card, gumming them down in their natural sequence. Leaving the pear-shaped eyes where they are, we commence with the first or internal antennæ, which we shall thus see has its basal joint spread out into a broad scale which ends in a sharp spine at the side of the second joint. It is in this basal joint that the Prawn's organ of hearing is contained. Spence-Bate and others have investi-gated this remarkable organ in various species, and find there is a little chamber with a slit-like opening, only to be seen when the creature has just shed his coat of mail, at which time he picks up with his "finger and

THE PRAWN.

thumb" a few minute grains of sand and carefully introduces
them into this auditory chamber, where they mix with some
fine hairs or cilia, and their agitation when acted upon by
sound-vibrations transmits sensation to the nerves The third
joint gives support to two lashes or "horns," one of which in
this species is branched.

The basal joint of the second antennæ also bears a flattened
scale that is enormous, being three-quarters of an inch in
length and a quarter of an inch in breadth.   There is but one
lash (*flagellum*) to this external antenna, and this strong, long
organ measures over six inches in full-grown specimens; that
is, one and a half times the extreme length of the Prawn from
the tip of the forbidding rostrum to the extremity of the tail.

We have mentioned the two pairs of "hands" (*chelæ*), and
behind these are three pairs of long and slender walking feet.
Then, further back, beneath the abdomen, there are five pairs
of swimming organs, and to these in the female the eggs are
attached. The tail-fan must not be forgotten; it is a beautiful
and most effective organ. The four plates of which it is composed are finely fringed with delicate hairs, and are so hinged
that they can be partially closed one over the other, or fully
expanded to have greater power when opposed to the water.
It is by means of this valuable organ that the Prawn takes
those astonishingly rapid backward leaps which make him
hard to be caught either by man or smaller enemies.

Before leaving the Prawn, I would like to say that our
portrait of him does not pretend to show the length of his
antennæ; and it would be well to make clear how he carries
so many to be useful to him. He is always waving these
about, and there can be little doubt that he receives impressions through their agency, olfactory and otherwise. It does
not matter how far away a Prawn may be; if you give an
anemone a small portion of food, and there is a Prawn at the
far end of the tank he will know it, and will come prancing up
to the right neighbourhood. But his olfactory sense though it

helps him to this extent, appears to act best at a little distance from the fragrant object. I have frequently observed a Prawn come quickly to the *locality* where food has been introduced and evince great excitement and interest; but his sense has not been fine enough to tell him at once the particular spot in the locality where it lay. I have on such occasions seen him walk over what he was seeking, whilst his hands were nervously scraping the ground and casting around for the delicacy he knew was close by. Now this is the order of his antennæ-bearing: of the first or internal antennæ that lash which has the short branch is carried half erect pointing outwardly, the companion lash pointing forwards, so that he cannot run against any obstruction without knowing it. The second or external antennæ are borne with a slight curve forward, then far abroad on either side. He is thus fairly guarded by sensitive organs well-nigh all round.

There are two other British species of *Leander*—*L. squilla* and *L. fabricii*—which occur in the rock-pools, and may easily be mistaken for young specimens of *L. serratus*. The distinguishing feature is to be found in that awe-inspiring, saw-edged rostrum that projects far in advance of the Prawn's head, and of which no one has yet discovered the purpose. In the *bona fide* Prawn this has a very decided curve upwards all the way, and on its upper edge it has seven sharp spines closely following each other, with an eighth lagging a sixth of an inch behind the seventh, and really on the carapace, not the rostrum; on the underside there are four close together in the middle, and a half-hearted one midway between the first of these and the tip of the rostrum. So much for the type; now for *L. squilla*. The rostrum is almost straight with a slight upward curve towards its tip. Like its big relative it has seven or eight teeth above, but *two* of these are really part of the carapace, and there are only *three* spines below. The second pincer-legs are not proportionately as robust as in the Prawn, and the creature does not attain to more than half the

Prawn's dimensions. *L. fabricii* has the rostrum nearly straight, with five teeth above and three beneath; in addition, this species has the rostrum covered with a multitude of minute reddish dots. There are similar dots in the Prawn, but none in *L. squilla*, with which it agrees more closely in size. These two, with young specimens of *L. serratus*, get mixed up and sold together under the name of Red Shrimps or Cup Shrimps.

There is a somewhat similar form called the Æsop Prawn (*Pandalus montagui*). It may be distinguished by the fact that the carapace is distinctly keeled along the foremost half of its upper part, and this keel is continued forward as the rostrum, which is armed above with moveable spines, while below it has five fixed teeth. The outer antennæ are long, and marked throughout by alternate light and dark bands. The inner antennæ have two lashes, the outer of the two thicker than its fellow. The first pair of legs are *not* furnished with nippers; and the second pair are very unequal in length and stoutness. It is reddish-grey in colour, dotted with a darker tint. Its length on our shores, according to Bell, does not exceed two and a half inches; but on the coasts of the United States it is said to attain to a length of four or five inches.

There is a beautiful little Crustacean, which may fitly be named the Varying Prawn (*Hippolyte varians*); it swarms in certain rock-pools and among the rocks at low-water. In such situations it is not so widely distributed as some of the species we have named, but it is worth looking for on account of its remarkable sensitiveness to the colour of its surroundings. Specimens taken from a pool in which the green *Ulva* or *Enteromorpha* is the prevailing vegetation are green; but if transferred to a vessel containing only brown, red, or yellow weeds, will in the course of a few hours be found to have changed their colours to harmonise with their new environment. So complete is this change that one can well understand how this shrimp may be commonly distributed all round our coasts, and yet only known from a few localities, because a careless observer would never see it. Like the species of *Leander* this has a rostrum—in this case quite straight, a sharp point. On the upper edge there are usually four teeth, but this number may be increased to five or even six; on the underside they never exceed two, and there may be only one.

Whilst referring to these little-known species of Prawns, we must not forget to mention the very well-known Common Shrimp or *the* Shrimp (*Crangon vulgaris*), which affects sandy shores and rivers rather than rocky coasts. The natural colour of the

COMMON SHRIMP.

Shrimp before it has been in the pot and made to reappear as the *Brown* Shrimp, is a pale brownish grey, thickly dotted with darker brown, which harmonises well with the sandy flats on which it loves to live. Looking at this species we see how great a finish is given to the Prawns by the possession of that saw-edged rostrum. By comparison the Shrimp has a square front, which is by no means so prepossessing. His eyes are not so distant one from the other as are those of the Prawn, and only one pair of his antennæ (the external) are at all long. There are three small spines on the carapace, one on the middle line and one on each side. The first pair of legs are stout, and what is technically described as *sub-chelate*, those of the Prawn's being *chelate*. The Shrimp's nippers have not got the well-formed moveable finger and fixed thumb of the Prawn, but a moveable finger and a little stump upon which it folds down. I do not pretend that it is not as efficient for the Shrimp's use as the better-looking contrivance of the Prawn. The plates of the tail-fan, too, are narrower than those of the Prawn, but the swimming feet are longer.

Now these two things would lead us to suppose that the Shrimp depends less on jumping back from danger than on swimming, and this is true. If the Shrimp suspects harm he sinks upon the sand, and setting his swimming feet rapidly to work they " kick up such a dust " in the water that he is hidden in a cloud of fine sand, which as quickly settles down and partially buries him—sufficiently so with his sandy hue to effectually hide him. Upon those swimming feet the female carries her eggs. From the fact that shrimps may be found laden with these eggs at almost all seasons, it would appear that they have no special breeding time; and this fact probably accounts for the endless supply of them. In common with most other small Crustacea they are constantly preyed upon by fishes, and we know something of the enormous mortality among them caused by man, when we think of the heaps in the fishmongers' shops and in the baskets of the itinerant vendors in towns. But the united efforts of man and fish do not appear to make them at all scarce.

There are quite a multitude of distinct species of British shrimps, but many of them keep away from the shore and are only caught in the dredge or the trawl. Some others swarm after the bait in lobster pots, though the lobster catcher does not want them, and does not even dignify them with a name— scarcely notices their existence, in fact. There remain, however, several species to which I must call attention, even though my readers may have expected me to have exhausted the list long before this.

The Chameleon Shrimp (*Mysis flexuosus*) will be found in summer to abound around the rocks and in the pools. It partakes somewhat of the character of *Hippolyte varians* in respect of colouring. If you take it around rocks that are covered with the *Laminaria* it is pale brown, or darker if from among *Fuci*, and in the pools where *Ulva*, *Enteromorpha*, and *Cladophora* prevail, its colour will be a light or dark green. It is a singular-looking shrimp on account of its long and slender carapace and the cylindrical abdomen. It has six pairs of feet, and not one among them all possesses a pair of pincers. The external antennæ are very long, and each is accompanied by a long flat scale similar to that of the Prawn's. The eyes are large and very prominent. The carapace is inclined to have a rostrum, but it is a poor attempt, and does not extend to more than a third of the eye-stalk. It is sometimes called Opossum Shrimp, because it has a peculiar pouch in which the eggs are retained until hatched, and where the young pass their early days.

There remain several species which should more fitly be included with the Lobsters, but from their small size they may pass muster with the Shrimps. They are exceedingly interesting, even if we take but one fact into account: their habit of burrowing in deep sand like mole-crickets. Right back in the early days of the present century an enthusiastic naturalist, Colonel Montagu, was digging for Razor-shells (*Solen*) in a sandbank near Kingsbridge in Devonshire, when he had the good fortune to turn up some things he was neither looking for nor suspecting the existence of—as a matter of fact they were quite unknown until Montagu unearthed them. Now here is encouragement for anybody and everybody who turns over weeds, pries into rock-pools and crannies, or digs in the sand for Launce or Razor-shells. You may or may not find what you seek, but something of interest you cannot help finding, and it may be a new fact—if not a new species.

When Montagu published a description of his find in 1808—

three years afterwards—it was under the three-barrelled name of *Cancer Astacus subterraneus;* but Dr. Leach, six years later, saw that it could not go into the same genus with the crabs or lobsters, and he called it *Callianassa subterranea,* by which name it has been known ever since. So far the account is plain sailing enough, but to attempt a description of *Callianassa* is not nearly so simple. The carapace is very small, with the slightest pretence to a rostrum, flattened at the sides, rounded above. The eyes very small, like those of its fellow-digger the mole, though more exposed than his. There are two pairs of antennæ; the internal ones double. The first legs are adorned with nippers, but they are very unequal in size, one being scarcely larger than the second or third feet, and the other much larger than the carapace, broad, flat, and hairy on the edges. On the outer side of the arm of this big limb there is a process which looks like a reaping hook. Now, the word *Callianassa,* I presume, is made up from two Greek words (Kalli, anassa), signifying Beautiful Queen; but I fancy that if a female monarch had one of her hands normal and the other bigger than her chest and head combined, none but courtiers would flatter her by declaring she was beautiful, and possibly they might be partly actuated thereby through a wholesome fear of that big hand. However, she is beautiful in respect of colouring—a fine bright pink, which departs with life. The second pair of legs are small and terminate in a little pair of pincers; the third have one finger which works against the enlarged next joint; the fourth terminate simply in a claw; the fifth in an intermediate condition as though the extremities intended to develop into pincers. The seven-jointed abdomen is long, the fifth segment broadest, from which it narrows gradually to the front, and suddenly to the rear, where it is finished off with a tail-fan of four plates. From a glance at the Beautiful Queen's hands and with knowledge of her burrowing habits, I should suppose that the bigger of the two served the double purpose of a digger and a street-door; the latter to keep enemies and prying intruders out of her burrow. Her majesty measures about two inches in length, and her crust is very thin and parchmenty.

That was a day to be remembered by Colonel Montagu, for on the same occasion he unearthed another burrower— *Upogebia stellata*—new to science. This is more lobster-like than Callianassa in form, though less so in size, for it is only about an inch and a half in length. It is content with having pincers to the first pair of legs, and these are nearly equal in size. All the limbs are liberally fringed with long hairs. The carapace begins with a small and sharp rostrum. Dr. Leach records it from mud in Plymouth Sound. Its colour is yellowish-white, sprinkled with minute orange spots.

And now, though we have by no means exhausted the list of British species, we must close this chapter. It should be stated that all these creatures go through a series of transformations similar to, but not identical with, those marking the early life of the crab and lobster.

CHAPTER XI.

SOME MINOR CRUSTACEANS.

BESIDES the crabs and shrimps already enumerated there are to be found upon our shores a great variety of smaller species of Crustacea, representing widely differing tribes and orders. We cannot fill a phial with water from a rock-pool without getting a number of specimens of the crystal-cased water-fleas (*Entomostraca*), of which we are probably already acquainted through several well-known fresh-water forms. We cannot pull up a tuft of fine weed from the same pool but we shall find on putting it into a tumbler of water that it harbours a multitude of Crustaceans much larger than the water-fleas; and so when we place in our aquarium a rough bit of rock, because it is the resting-place of a tube-worm, an acorn shell, or a patch of polyzoa, we shall find it is also occupied by little shrimp-like, or woodlouse-like creatures. There is every probability, too, that we shall get with these the minute larval forms of crabs and lobsters. It is a delight to introduce them in this way, and to be constantly making the acquaintance of unsuspected inmates of an aquarium that perhaps only holds a couple of quarts of water.

Of course, there is no difficulty in collecting these smaller species of set purpose, any more than there is in looking for anemones and sponges; but whether the shore naturalist seeks them or not, he is bound to get a large variety.

The majority of these will be species of the two important sub-orders, Isopoda and Amphipoda, and one of the most conspicuous, because largest, of them is the Sea-slater (*Ligia oceanica*), represented in our next illustration. It will be found crawling up the perpendicular faces of rocks about half-tide mark; and the finder will not need to have explained to him the fact that it is related to the terrestrial Woodlouse or Slater of our hedgebanks. The whole tribe have the respiratory appa- ratus adapted for breathing air, but they appear to require a damp atmosphere.

Among the fringing weeds of the rocks there will be found great numbers of a lively creature of some- what similar build to the *Ligia*, but very narrow (oblong-ovate is the tech- nical description), and with- out the terminal appendages (*uropods*) of that creature. It varies in colour from pale-brown to a dark-brown, perhaps mottled with black. There are several British species, but the common shore-haunting kind is *Idotea marina*. Its great variation has caused it to be called by at least a dozen names.

In turning over any or- ganic remains above the reach of the waves, we shall uncover swarms of the Shore-hopper (*Orchestia littorea*), distinguished from the similar Sand-hopper (*Talitrus locusta*) by its more compressed body, and by having both the first and second pairs of feet clawed, whereas in *Talitrus* the second pair are not clawed.

SEA SLATER.

Among the dried up, black-looking foliage of *Lichina pygmæa*, which grows on the rocks that are covered only for a short time at high-water, will be found the queer Isopod, *Campecopea hirsuta*, which seems to mimic the plant that shelters it. They curl up tightly into a ball, and roll about if dislodged. The projections at the end of the body (*uropods*) help their resemblance to the *Lichina*. This species must not be confounded with the similar and allied *Næsa bidentata*, which has the sixth segment of the trunk much larger than the others, and produced backwards in the *two teeth*-like processes, which suggested its Latin name.

If one is so fortunate as to get access to the rocks at the equinoctial low tides, which are lower than the ordinary fortnightly "springs," he will see rocks covered with a muddy felt, much of which appears to be the work of marine worms, who live in it. A portion of this coating should be rapidly prised off with the putty-knife, and put into a bottle of sea-water by itself. At the same time look for a dirty-looking slaty rock, at the same level, and take off the upper flakes, with their investing crust of acorn-shells, corallines, zoophytes, etc. On this will almost certainly be found the absurd acrobat or contortionist, the Skeleton-shrimp (*Caprella linearis*), sprawling about, his walking-feet on the extreme segments of an extremely long and thread-like body. Here will, in all probability, also be found a Crustacean with a body not more than half an inch long, but looking much longer by reason of an enormous development of its outer antennæ, which it flourishes about as though they were long arms. The chief use it makes of these is as flails to thresh out its prey, certain marine worms that inhabit the mud-felt to which we have referred. By repeated heavy beatings on the mud with these antennæ, the worms are induced to come outside their burrows to see what danger is threatening them, and find out only too quickly.

The first time I saw this remarkable creature, I was greatly moved to mirth. I had wrested a flake of rock from a huge mass that was ordinarily covered at low-water, but which now

at the equinox reared its head high above the waves, and
exposed treasures in the shape of the Globehorn and the
Rosy Anemones. *Corynactis* was growing at the edge of this
flake, which was placed near the glass of a small aquarium,
where it could be easily scanned with a lens. A few hours
later I took a glance at my Globehorns, and was astonished to
witness the activity and vigour of the varied colony that was
settled on these few square inches of stone. Several acorn-
shells were in "full swing," a tube-worm (*Sabella*) had put out
its plumes from the mouth of its tube, a patch of polyzoa
exhibited its crowns of prismatic tentacles, a couple of *Caprellæ*
were sprawling around in an inebriated fashion, whilst near
one corner was the figure that chiefly attracted my attention.
Corophium longicorne was standing erect in a mud-pulpit,
above the walls of which he was flourishing his arm-like
antennæ as he—a Crustacean St. Anthony—harangued the
other members of the community who appeared to be paying
great attention to his discourse. I felt that if I could but
restrain my laughter, I should hear the "thirdly, my brethren
beloved," and the telling sentence he emphasised by a hearty
smack on the pulpit; the ridiculous *Caprellæ* profoundly bow-
ing in assent to his postulates all the time.

CHAPTER XII.

BARNACLES AND ACORN-SHELLS.

Occasionally in strolling along a beach after a storm we shall encounter some wreckage that came ashore with the last wave of the incoming tide, and so failed to be washed off again. It may be a spar, a rudder, a stern-board with a name upon it that tells a tale of a vessel that has gone down. It may come in clean, with the splintered wood looking as though just smashed, and we may judge from such appearances how long it is since the catastrophe happened. On the other hand, it may bear evidence of having floated in the sea for a long period before getting into a current running coastwards. Such evidence will consist in the wood being heavily soaked with water, or in its surface being covered with hundreds of writhing snake-like creatures with pale-blue heads. We have met under such circumstances, with balks of timber with scarcely an inch of their surface not covered with this foreign growth; with casks on which they grew all round the edges of the heads and the hoops.

A few months ago there drifted into our "porth" a small keg-buoy with a long thick hawser attached, and the submerged half of the buoy had a fine crop of the writhing things hanging from it, whilst they hung from the rope in clusters a few inches apart. The finder very kindly hauled it upon the rocks, and coiled the hawser round it that I might photograph the entire lot. As it lay there in the autumn sunshine it looked a very pretty group, and I regret that the camera would not reproduce the snaky movements, nor the fine colouring.

Now the creature is no other than the Ship-Barnacle (*Lepas anatifera*), one of the chief obstacles to speed in the old days of "the wooden walls of England." When a ship had made an ocean voyage it was necessary to dock her and scrape off the enormous quantities of Barnacles that not merely added to her weight, but offered strong opposition to her passage through the waters. To-day, what with steel vessels and patent anti-fouling compositions with which to paint the ship's

SHIP-BARNACLE.

bottom, the poor Barnacles find their world much narrower than formerly, and with fewer openings for the enterprise of their race. Should you come across such a barnacle-ridden waif of the sea, consider it carefully. You shall find in it matter of interest, and, in addition to its provision of something for your imagination to play round, in your efforts to get a clue to the vessel of which the wreckage once formed part, the life-story of the Barnacle itself is a romance.

Before we attempt to tell this story briefly, let us look at one of the specimens before us. The long and evidently muscular neck ends in a composite shell, which is seen to be composed of four portions, or valves hinged together, opening in front, and strengthened at the back by a fifth valve, a long, narrow, and curved piece. At short intervals the two halves into which this "shell" is obviously divided part in front, and out comes a mass of coiled up, slender, and hairy processes which separate and uncoil as though attempting to catch some invisible body, then coil up again and withdraw as though they had really caught it and meant to keep it. Now this is the principal, one might almost say the sole occupation of their adult lives, but writhing is another to which they pay some attention. Probably it may strike you as a monotonous, perhaps senseless way of spending one's days; but it is quite evident, from the great numbers of Barnacles crowded within a few square feet, and all looking prosperous, that it is a paying game.

It must be remembered that however clear and crystalline the sea-water appears, there is really great truth in the remark of the scientific luminary, who said that the sea was a kind of thin soup or broth, holding enormous quantities of animal and vegetable matter in solution, most of it invisible to the unassisted vision. Whoever possesses a retentive hand like that of the Barnacle, has only to spread the palms and fingers wide, then close them tightly, to have *something* enclosed therein. Such is the Barnacle's experience; and it is by the mere opening and shutting of his hand that he gets a good living. Strictly speaking, this hand is *not* his hand, but a number of feet and hands which correspond with the limbs of the crabs, lobsters, and shrimps.

Strange as the assertion may sound, unlike as the creatures appear, the Barnacles belong to the same great class (Crustacea) as the animals described in the last two chapters, though they are partly separated from them and put into an order (Cirripedia) by themselves. No wonder if you hesitate to accept this statement as a fact; you are in good company, for no less a naturalist than the great Cuvier failed to see the relationship.

That this order is an important one will appear when it is stated that the great Charles Darwin wrote an important work in two volumes, devoted to the "recent" Cirripedes, and two other volumes on the "fossil" species of the order.

These Cirripedes are divided into two main groups—the pedunculated or stalked Cirripedes, represented by the lively Barnacles before us, and the sessile or stalkless Cirripedes, of which the familiar Acorn-shell of the littoral rocks are the examples.   .

Now these two groups may strike you as having little in common, and yet their early history is practically identical, one group with the other. Longfellow was quite right when he stated that "things are not what they seem," at least, they are not *always* what they seem; conversely, they do not always seem what they are. We must not be content with taking a couple of creatures at one particular stage in their existence, and say these organisms differ so widely from each other that we must put them into equally widely separated classes or groups; we must try to find out and compare all the stages in their life-histories, before we can talk of separating or bringing together, except in the most temporary fashion, there to be kept, as it were, in quarantine until we have found out what we wish to know concerning their antecedents.

No one, until he had evidence of the successive stages in the life of a butterfly, would dream of putting such dissimilar things as a caterpillar and a butterfly into the same order; yet their wonderful course of development was long ago traced out, and it is within the power of any person to check off the whole progress from the batch of elegant eggs laid on a cabbage leaf, through the ravenous worm-like caterpillar stage, and the apparently inanimate chrysalis to the beautiful white

butterfly that can take no solid food, and which by depositing
another batch of exactly similar eggs, completes the cycle,
and so assures us we have made no mistakes in our observations.

In a like manner we can watch the series of stages, utterly
unlike each other, through which a crab, a lobster, a shrimp
or a Barnacle passes before it attains the adult condition ;
and when we find the early forms of the Barnacle agreeing in
a very curious way with stages in the life-history of typical
Crustaceans, we are perfectly justified in grouping them in
the same class of animal life.   We have, in fact, pierced
through the disguise with which some of the adult forms have
sought to hide their identity, and have found out their true
characters.

It must be confessed that the course of development in
some of these creatures partakes of the character of what has
been termed " an Irishman's rise."   In the case of the caterpillar and the butterfly, everybody recognises that development is progress, that the butterfly is a higher being than the
caterpillar.   But in others development spells retrogression.
Such is undoubtedly the case with the Cirripedes, and with
certain crustaceans which lead the life of parasites.   The
course of development in the Barnacles and Acorn-shells has
been very succinctly stated by Darwin.

" The larvæ in the first stage have three pairs of locomotive
organs, a simple single eye, and a prosciformed mouth,
with which they feed largely, for they increase much in size.
In the second stage, answering to the chrysalis stage of butter-
flies, they have six pairs of beautifully constructed natatory
legs, a pair of magnificent compound eyes, and extremely
complex antennæ; but they have a closed and imperfect
mouth, and cannot feed: their function at this stage is to
search out by their well-developed organs of sense, and to
reach by their active powers of swimming, a proper place
on which to become attached and to undergo their final

metamorphosis. When this is completed they are fixed for life: their legs are now converted into prehensile organs; they again obtain a well-constructed mouth; but they have no antennæ, and their two eyes are now re-converted into a minute, single, simple eye-spot. In this last and complete state, Cirripedes may be considered as either more highly or more lowly organized than they were in the larval condition. But in some genera the larvæ become developed into herma-phrodites, having the ordinary structure, and into what I have called complemental males; and in the latter the development has assuredly been retrograde, for the male is a mere sack, which lives for a short time, and is destitute of mouth, stomach, and every other organ of importance, excepting those for reproduction."

In this early condition these Cirripedes much resembled the minute so-called water-fleas that swarm in our fresh-water ponds and streams, and when upon the point of their last change they laid their heads down upon the spot selected for their future station in life. Then a natural marine glue, that sets under water, exuded from their antennæ, and they became fixtures, head downwards. The two valves of their old shells were thrown off, and the new ones, largely composed of carbonate of lime, grew up from the base.

Some of the Barnacles on our buoy are apparently dead, and one of these we can take to pieces. Taking off one half of the compound shell, we find the creature attached to the floor of the chamber, evidently on its back. From the upper end there arise the twelve limbs, six on each side, and each one dividing into two branches, each branch a beautiful feather with a wonderfully jointed, supple, purple-black stem, closely fringed with purple hairs. It is from this plume-like cluster of curling limbs that the order obtains its name (Latin, *cirrus*, a curled lock of hair, and *pes*, a foot=curl-footed).

When the shell opens and the trunk which supports all these limbs is thrust forward, each branch separates from its

fellows and becomes almost straight, spreading out its hairs
as widely as possible. Thus extended, the entire plume of
feathers sweeps through a limited space of water, and many
minute creatures are entangled in its hairs, and so brought
into the currents that flow towards the Barnacle's mouth.

Huxley has described the Barnacle as standing on its head
and kicking food into its mouth; but we question whether this
partakes of his usual accuracy of description. So far as we
have been able to make out the process, the food particles are
strained off from the sea-water by this exquisite net, and
brought, not kicked to the mouth.

It is to this plume of feathers that the Barnacle owes its
specific name, *anatifera* = goose-bearing. It was formerly
thought to be a vegetable production, whose fruit, when ripe,
gaped open, and dropped out an embryo bird, which fell into
the water and developed into a Bernicle Goose. Gerarde,
three centuries ago, wrote a wonderful and circumstantial
account of the whole business, which he declared he had seen
with his own eyes; and every writer of popular works on the
sea since then has seen fit to reproduce his account as one of
the curiosities of natural history. I have no intention of doing
so, for it is time it had a little rest after being so hard worked.
For a similar reason I have in this book utterly ignored
Montgomery's " Pelican Island; " and the equally hackneyed
quotations from Southey, Crabbe, and Coleridge, that have
been a boon to some of my predecessors in filling their pages,
I have also put upon a retired list.

Cirripedes, not being so completely boxed up as the majority
of crustaceans, can enlarge their dwellings by additions to the
edges of the shells, and therefore do not need to throw off
the entire envelope from time to time.   But it is difficult to
entirely get rid of racial characteristics, even when there is no
special need to retain them; and so we find the Cirripedes
casting the skins of their bodies from time to time, though the
limy shell is made to serve for all their life.

There is a smaller species of Necked Barnacle (*Scalpellum vulgare*), the shelly portion of which, seen edgeways, looks like a penknife, whence the Latin name. It is usually found growing among corallines; it is figured in accompanying group.

PYRGOMA. SCALPELLUM. PORCATE-BARNACLE.
ACORN-SHELL.

There is a peculiar little Barnacle called *Pyrgoma anglicum*, which is parasitical upon the pretty Devonshire Cup-coral (*Caryophyllia smithii*). It is shown on the coral in the upper left-hand side of our illustration above, and may be looked for in any of the localities where this coral occurs. It attaches itself to the outer edge of the plates of the corallum.

Let us turn now to the more familiar Acorn-shells (*Balanus balanoides*) that crust the rocks between tide-marks. We might have used the expression "too-familiar," for whoever has had to put a bare-foot upon them in bathing or swimming from the rocks, will have had cause for remembering their sharp edges. It is not easy to keep the Ship-barnacle in an aquarium; but a flake of rock, or a disused limpet shell, crusted with *Balanus*, is conveniently kept in a glass of sea-water, and will long continue at once a thing of beauty and a wonder to friends who are ignorant of natural history. These are sessile Cirripedes, that is, they have no stalks upon which to writhe, but *sit* directly upon the rock.

If we scrape one of these Acorn-shells off the rock with our useful putty-knife, we shall find that it has a thin base of shelly matter upon which it reposes much as the Ship-barnacle does upon the floor of its shelly chamber. But it will be seen that the sloping outer walls of the Acorn-shell are firmly cemented together, and allow of no movement; the top, however, is open, but the animal within is protected by an interior door of four pieces, that opens in the middle like the cellar-flaps seen in connection with business basements. These doors "butt" together accurately, and open easily by pressure from inside. Then out comes a more beautiful and delicate "hand" even than that of the Barnacle, for this is so fine and transparent that it looks a thing of spun glass. There is the same movement as in the Barnacle, the everlasting grasping at something, the opening and shutting of the cellar flaps. Its earlier history is also similar to that of its stalked relation. There is a larger species of Acorn-shell known as the Porcate-barnacle (*Balanus porcatus*), the name having relation to the form of the conical shell; porcate signifying that it has ridges between the furrows that mark its outside. Other species, smaller, some almost flat, will be found on some parts of our coast, but we would refer our readers to Mr. Darwin's work* for the further study of the Cirripedes.

* A Monograph of the Cirripedia, 2 vols. Ray Society.

CHAPTER XIII.

"SHELL-FISH."

ONE of the greatest hindrances to the unscientific, in the way of a proper understanding of the true nature and relative position of many forms of life, is to be found in our misuse of words—our poverty of language, which compels us to make one word serve for quite dissimilar and unrelated things. This unfortunate term, "Shell-fish," which we have felt compelled to put at the head of this chapter, in place of the more accurate "Bivalve Mollusks," is a case in point. I really want a name that only includes these; but in order to be strictly popular in my chapter-heads, I must use this very general term. Just now I turned to a popular and portable dictionary to see what was a familiar definition of the compound, and I read there, "Shell-fish, testaceous mollusks," but even for a popular explanation that does not go far enough, for Shell-fish also includes crabs and lobsters, which are not mollusks, but crustaceans. I daresay, too, that in a fishery suit, if it served their purpose, lawyers would show plainly that it embraced tortoises and turtles, which are chelonian reptiles. We are all aware that in popular and legal language everything that comes out of the sea is a fish, excepting the coral-polyp which everybody, except naturalists, knows is an insect!

What I really wish to make clear, after this little growl, is that the present chapter will deal only with such creatures as are (like oysters and cockles) sandwiched or boxed between two valves or half-shells, and will not even glance at those mollusks that are contented with a shell all in one piece; these are relegated to the next chapter.

The Mollusca that actually live between tide-marks, though numerous as individuals, do not represent many species; but those of which we may find the recently-vacated shells, thrown up by the tide from greater depths, will total up to a considerable number. The bivalves must be sought for on sandy

SPINY COCKLE.   BANDED VENUS.

beaches and mud flats, especially at the mouths of rivers. Most of them are burrowers, excavating a way by means of the powerful foot with which they are provided. This instrument is well seen in the Razor-shells (*Solen*), or the Cockles (*Cardium*), where it reaches extraordinary development. Even where the animal lives far beyond our limits in deep water we can, by a little thought, get some notion of their habits by examining the empty shells that are cast up within the littoral zone by heavy seas. Those that are fresh and clean externally, though without any signs of wear from long washing among the shingle, may be safely regarded as burrowers that habitu- ally lie beneath the sand or mud. These, too, will be found to have both *valves* of the shell almost, if not quite, equal in size and shape; whilst those which, like the Oyster and the Scallop, lie upon the sea-bottom, have very unequal valves, the under one being deeper and concave, whilst the upper valve is flat and more brightly coloured, to harmonise with its surroundings. Often, too, this exposed upper valve will be crusted with acorn-shells, *Serpulæ*, *Sertulariæ*, or seaweeds.

It may prevent confusion further on if we now say a few words by way of defining the parts of a bivalve shell, its latitude and longitude, and its relation to the animal whose vital activi- ties produced the valves. The Spiny Cockle, or Red-nose (*Cardium aculeatum*) of our illustration, is on its back. If we were to take it, or any other bivalve-shell, and turn it the other way, so that the hinge connecting the two valves was uppermost, we should have it in the natural position.

A bivalve mollusk is an inferior creature to a limpet or a winkle, because these have heads with eyes, but the bivalve has not. In the larval condition it has eyes, but by a retrograde movement like that of the cirripedes, it gets rid of these as useless in the life it is to live henceforth. But in spite of its want of a head, we know which is its anterior and its posterior end, its dorsal and its ventral surface; and with our know- ledge of the relation of animal and shell, we are not troubled to open the valve to look at the creature, when we wish to describe the parts of a shell. It will be noticed that each valve curls over near the hinge and takes a form not greatly unlike a beak. This is more strongly marked in some species than in others; anyhow, it is popularly known as the beak, though it is technically distinguished as the *umbone*, or boss.

If these beaks have the slightest tendency to either end of
the shell, it will be to the front, where we should expect the
creature's head to be, if it had one. This point made clear,
by reference to the shell we have just picked up, we can say
which is the right and which the left valve. The valves are
hinged by a band of a substance that looks much like catgut.
It is elastic in character, and is always pulling at both valves,
so that the natural tendency of the shell is to gape open. But
inside the shell there are, in most bivalves, two much more
powerful bands of muscular fibres (the oyster has but one),
which, by their tension, can slowly or suddenly bring the
edges of both valves closely and tightly together, and hold
them so for an indefinite period. You can see the marks
where these muscles were attached, one at each end of the
valve. Between these two marks ("muscular impressions")
there runs a colourless line marking the area to which the
mantle was attached ("pallial impression"), but this line is
often interrupted, towards the hinder end of the shell, by a
bay or sinus (the "pallial sinus").

The mantle is a delicate membrane on each side of the
mollusk's body, which has the power of forming the shell, to
which it is attached save at the margins. The "pallial sinus"
is caused by the syphons which are protruded at that end of
the shell. At the other end, as shown in the figure of the
Banded Venus, is the "foot." The "syphons" are two delicate tubes, and if you were to put a living Venus, or other
syphon-bearing mollusk into a glass of clear sea-water, and
drop a little finely-divided indigo, or other colouring matter,
in the immediate neighbourhood of these syphons, you would
observe a stream of the minute colour-particles rushing into
one of these tubes, and a stream of clear water issuing from
the other. The inflowing stream passes between the leaf-like
gills, or respiratory organs ("branchiæ"), where it is effectually
strained, all solid matter being retained and passed on to the
stomach, whilst the filtered water passes out through the

second syphon. The length and form of these syphons differ in distinct species, but each kind is pretty true to its own type, and, consequently, the impression that it makes on the interior of the shell, taken in conjunction with the muscular and pallial impressions and the hinge-teeth, are a certain guide to the discrimination of species.

These are matters that are essential to one's knowledge of the mollusca, and they must be learnt; but the few species we shall be able to mention in this chapter will be indicated more by their external shape, marks, and colouring. When so identified, the reader should strengthen his knowledge by a practical study of these internal impressions, and the characters of hinge and teeth.

This Spiny Cockle, or Red Nose (*Cardium aculeatum*), is not the Vulgar Cockle (*C. edule*), although it is much sought for food on its native Devonshire coasts. It is a very much larger species than the last-named, and gets its name of Red Nose from the brilliant hue of its long strong foot, which is at once a burrowing instrument and a leaping pole. By pushing its pointed end down into the sand, and then bending it into a hook, it can, by contracting the foot, pull the thick prickly shell down after it. On the other hand, by pressing its bent tip against some unyielding substance, it can use it as a spring, which shall suddenly send the shell flying through the water to some considerable distance. The Spiny Cockle is a creature of clean, sandy beaches, where it may be found at low-water, but only on the Devonshire coast.

The Common Cockle (*C. edule*) is very much smaller, its shell free from prickles, and marked merely with bold rounded ridges. It is more likely to be found where the sands are not wholly of sand, but contain a liberal admixture of mud. On some of our coasts it is exceedingly abundant, and in times of famine has saved populations from starvation. It is certainly on record that the people of the Isle of Barra, in the Hebrides, have been thus preserved many years ago, when all the people sought the Cockle on the great expanse of sands at the northern end of the island. "It was computed that for a couple of summers, at the time alluded to, no less than from one to two hundred horse-loads were taken at low-water, every day of the spring-tides, during the months of May, June, July, and August."

The Cockles have gained their name of *Cardium* and Cardiaceæ from the fact that if the shell is viewed "end on"—the curving beaks, of course, uppermost—it will present the conventional heart-shape (*Kardia*, Greek—heart). Some nearly allied genera, exhibit a similar form, but narrower, and therefore not so suggestive of hearts; but the Heart Cockle (*Isocardia cor*) is more truly heart-shaped than the Cockles of the genus *Cardium*. It is about three inches across its longest diameter, very thick and heavy, and the beaks are so greatly curled that no one will be disposed to quarrel with the name, either of the genus or the species. It is a deep-water species, but in suitable localities the empty shell may be found washed in by gales. It is chiefly found on the west coast, and it is probable that its headquarters, in British waters, is around the Isle of Man.

Several of our most familiar bivalves are not very distantly related to the heart shells. There are, for instance, the Venus shells of which we have already given a figure of one species, the Banded Venus (*Venus fasciata*). It is a solid, heavy little shell, of some shade of brown, with broad bands of a lighter hue radiating from the beak. A series of strong ridges run parallel with the margins, or, as usually expressed, the ridges are concentric. The various species of the genus inhabit sand and gravel from low-water mark to a hundred and forty fathoms. The animal must be obtained by the dredge, but the empty shells are thrown up freely after storms. A much larger species is :—

The Warted Venus (*V. verrucosa*), a drab-coloured shell, with very rough and unequal ridges. In some specimens these ridges are so broken by radiating lines, that the ornamentation has the appearance of being warty. The various species of Venus have three strong hinge-teeth on each valve, and the inner edge, though at first sight smooth, is very finely " milled."

The finest of these shells is the large, heavy Smooth Venus (*Cytherea chione*). It is a deep-water species, found chiefly on the southern and western coasts, where, in spite of its great weight, it is frequently washed up after storms. It is wonderfully smooth, inside and out; even the lines of growth are not high enough or sharp enough to take off this smoothness of the outside, which is coloured of a pale pinkish-brown tint marked by concentric lines of a lighter hue, and by much darker radiating bands. It is all very simple, but very effective. The inside is coated with white, and the muscular and pallial impressions are very strongly marked, though in no way interfering with the general plan of entire smoothness. The edges, too, are rounded and as smooth as the edge of a tea-cup. It is three and a half inches across the longest diameter of the shell, and its circumference, at right angles to the last measurement, is eight inches. The hinge-ligament is an inch long, and the teeth are very strong and prominent.

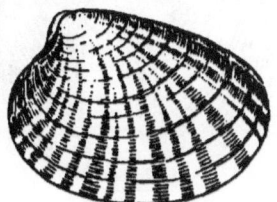

SMOOTH VENUS.

It is by no means a common shell outside the districts mentioned, but I have frequently found it on sandy shores in Cornwall, thrown up by storms, with the living animal still within. I believe most of the entire shells found on beaches have been thrown up whilst the animal is still in possession, and evidence upon this point may be obtained by examining the ends of the shells. It will be found that those which came to the surface with the animal are more or less chipped at the ends, where a Puffin, or other sea-bird, has cracked off a portion to enable it to prize the valves open; additional evidence will be found in portions of the muscular bands still adhering to the valves.

N

On the same sand and pebble beaches we shall find in greater plenty another of the Venus shells, the Rayed Artemis (*Artemis exoleta*). We presume that Linnæus, in giving this

species its name of *exoleta* (Latin, worn-out), was struck by the fact that however fresh a specimen may be, it has the appearance of having been knocking about with sand and shingle for some time. The shells are white, with variable rays of pinky-brown (sometimes entirely absent), and finely and evenly marked with concentric grooves. In proportion to its size, it is a very thick shell;

RAYED ARTEMIS.

very round in outline, except that a piece appears to have been nicked out of the edge in front of the beak. When the shell is closed, these marks on the two valves, coming together, form a heart-shaped depression of a brown tint, and called the *lunule*.

The lunule is not peculiar to this species, but is shared by a large number of bivalves. It is well-marked in the Smooth Venus, but not so completely heart-shaped as in the Rayed Artemis. There is a finely-developed set of hinge-teeth, and the pallial impression is deeply sunk. A closely allied species, the Smooth Artemis (*A. lincta*), is smaller, not banded, and the concentric ridges are finer and less perceptible. It is this peculiar type of smoothness that suggested the specific name of *lincta* (Latin, sucked), its appearance being as though a specimen of *exoleta* had been sucked until smooth. Both these have a hatchet-shaped foot for digging into the sand. Great quantities of this bivalve are washed up in winter, and I have frequently come across a piece of rock protruding through the sand, around which there were dozens of these shells, broken or chipped, giving evidence, from their fresh muscles, that they had but recently been destoyed. It has reminded me of the favourite stone under the hedge, whereto the Thrush brings her snails to be hammered until the shell yields up its luscious contents. Artemis has met with a fate similar to that of the hedgerow snails, for her fortress has been broken by gulls, puffins, or even by ravens when winter has taught them not to be too particular about their food.

There is a group of Venus-shells whose exterior is ornamented with concentric and rayed rounded ridges, in some cases strong, though regular and even, whilst in others they are but slightly perceptible; but their place is, to some extent, taken by colour. They bear the generic name of *Tapes* (Latin, tapestry) which is exceedingly appropriate, for the patterns of some species is very suggestive of tapestry and carpet. Especially is this so with the Cross-cut Carpet-shell (*Tapes decussata*), whose exterior looks like the back of a piece of tapestry carpet, both in texture and colour. The latter is of a nondescript drab, with occasional tinges

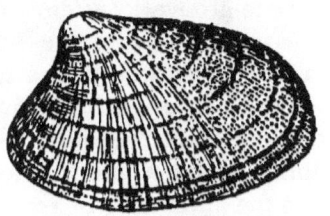

CROSS-CUT CARPET-SHELL.

of red and stain-like smears of bluish-grey. The ridges radiate from the beak, and they are nicely rounded, but their lines are by no means straight. They are cut across by fine concentric incised lines, which, with the grooves beside each rayed ridge, produce the cross-cut, or *decussate* appearance which suggested the name. The interior is dull white, like the surface of an enamelled card, the muscular and pallial impressions polished, and consequently very obvious.

The Virgin's Carpet-shell (*T. virginea*) is smaller, the exterior very smooth, the ornamentation taking the form of shallow concentric lines. The colouration is a mottling of salmon-pink, with little specks of white showing through, and irregularly-shaped spots of dark-brown sprinkled sparingly over all. Viewed not too closely, it will be seen that the whole surface is divided between about six broad rays of lighter and darker tints. The interior is white and glossy, deepening to pink or yellow towards the beak and hinge.

The Golden Carpet-shell (*Tapes aurea*) is similar, but some shade of yellow takes the place of the pink in the last-mentioned species.

The Pullet Carpet-shell (*T. pullastra*) is broader from the hinge to the edge of the shell, in proportion to its length at right angles with that measurement. Its name, *pullastra* (Latin, a pullet), has evidently a relation to its colouring, which is similar to that of *virginea*, though darker. If the two are compared it will be found there is a further difference in the fact that whilst *virginea* can scarcely be said to have any radiate-grooves, *pullastra* is covered with them; but they are not appreciable to the sense of touch, and scarcely so to ordinary eyesight, unless special attention is drawn to them— they are so exceedingly finely cut. Inside, the shell is white, that part lying between the impressions and the hinge being dull like the whole interior of *decussata;* but the impressions and the outer margin are polished.

The Tapes animals spin a byssus like the Mussel; they burrow in the sand at low-water with their thick fleshy foot, or spin up to the roots of *Laminaria* and other seaweeds among the low-water rocks. Around the shores of the European continent they are used as food, but do not appear to be so utilised in Britain.

The Scallops are familiar to all in a general way, from the frequency with which one species occurs on the fishmongers' slabs. This is the largest British species, and is generally distinguished as the Common Scallop, Quin or Queen (*Pecten opercularis*), a deep-water species, whose valves are frequently washed up on the beach. They occur in beds, but are not fixed like the Oyster; on the contrary, by the sudden closing of their valves and the consequent rapid expulsion of water, the shell shoots hinge foremost through the water to some considerable distance. The young ones can attach themselves by a byssus to the rocks, as is done by the Mussels, Carpet-shells, and others. It is a peculiarity of the Pectens that they have a pair of " ears " to the shell, the edges of which afford a good foundation for the hinge ligament, whilst in lieu of hinge-teeth to keep the valves firmly together when closed by the muscles, the corrugations of the valves extend right to the

COMMON SCALLOPS.

margins, and the ridges of the right valve fit into the furrows of the left valve and *vice versâ*. It will be noticed that these ears are not a good pair—one is always larger than the other, and the smaller one is popularly supposed to have been broken; that, however, is a mistake, the Pectens are built that way. The most prominent ear is always the front one, and below that of the right valve there is a notch where the byssal threads issue. In the Common Pecten these ears are much more nearly equal than in the others, whilst in *P. varius* there is a great contrast in the size and shape of the two, and in *P. tigrinus* one is almost absent altogether.

The Common Pecten is sometimes dredged for, but as a

rule it is avoided by the fishermen, on account of the risk to their nets and the small price realised for the mollusks after they have caught them. It will be noted more conspicuously in this species, on account of its size, that the right valve, which is the lower one when the creature is lying on its bed, is far more convex than the left or upper one. It is exceedingly variable in colour.

The Variable Pecten (*P. varius*) carries out its name to the letter, for out of a score one could scarcely find two that agreed in colour and the disposition of the markings. Their usual tint will be found among the red series of the chromatic scale; sometimes almost white with dark-red blotches, at other times dark red-brown with faintly perceived cloudings of a still darker hue. The exterior is ornamented by about twenty-eight bold ridges radiating from the sharp beak, and each of these, as it approaches towards the other edge of the shell, gives off irregular spiny processes. There is a rare variety of this

SCALLOP HUNG UP.

almost entirely white. *P. varius* is not content with using its byssus only in the days of its youth, but continues to do so, even when at full age; it may sometimes be found thus hung up to a rock, as shown in our illustration, or attached to the roots of *Laminaria*.

A live Scallop of any species, in a glass vessel of sea-water, is a beautiful object. It will soon open its valves and exhibit the richly-frilled edges of its brightly-coloured mantle; this organ, in fact, has a double margin, the inner of the two finely fringed, and at its base a row of eye-like beads.

When prying curiously about the rocks at low-water, under the scrub of weeds and corallines, we are sure to discover little flat pearly shells which we shall almost as surely decide to be young oysters. They are a kind of oyster, though not edible oysters of the genus *Ostrea*, but Saddle Oysters, of the genus *Anomia* (*A. ephippium*). Although they appear to be firmly cemented to the rock by the lower (right) valve, this is not really so. The thin blade of a penknife gently pushed beneath will move it off with the merest touch, for instead of being fixed by its whole under-surface, it is merely attached by a muscular plug that passes through a comparatively large oval hole in the under shell, near the hinge, and sticks like a sucker to the rock. As it grows older it will probably alter its form, to adapt itself to things it comes in contact with, as its diameter increases. Small specimens, not so large as a three-penny-bit, usually have a colony of much younger individuals located on their upper shell.

Odd specimens of the Common Mussel (*Mytilus edulis*) will be found among the roots of weeds on the low-water rocks, but to obtain them in quantity one must go to a mud-bank, as at the mouth of a river; or they may be found clinging in masses to the wooden piles of piers and breakwaters by means of their byssus-threads. The first thing a mussel does on being placed in an aquarium is to attach itself to the side by this means. Possibly he will wander a little, by means of his foot, to make sure of the right spot upon which to cast anchor, but having settled that point and found that he has made the right choice, there he will remain.

A mussel is the best of all bivalves for aquarium life. It is true he is not very lively, and does not flit through the water like the young Scallops. He is anchored, and there he stays, simply opening his shell a little way and putting out the frilled edges of his mantle, with their openings for the inward and outward currents—the inward bringing both oxygen and food, the outward carrying off carbonic acid gas and other waste.

The large, thick, coarse-looking mussel shells we occasionally find on the sands, measuring five or six inches in length, belong to a different genus, and are called Horse Mussels (*Modiola modiolus*). The valves in question may not have come far, for the species occurs in sand and mud as near as low-water. But it will not be found moored to rocks and weeds by its byssus; instead, it burrows and weaves its enormous byssus into a nest with sand and gravel mixed up with the threads. They are said to be coarse and unpleasant tasting, so that they are not used as food, and hence the name, Horse-mussel; the prefix, horse or dog, before a popular name for animals or plants, denoting its worthlessness as food, the sole criterion of worth to the popular imagination being found in the answer to the query, " Is it good to eat?"

A very handsome shell, as well as a common one, is the Comb-shell (*Pectunculus glycimeris*), whose thick round valves may be found rolling on the beach, where they have been washed up from the zoophyte ground in deep-water. It is very variable in its mark- ings, and yet there is so strong a family likeness running through all its variations, that there is not the slightest difficulty in its identification.

COMB-SHELL.

To the touch the exterior is quite smooth, though not glossy, but examined with a lens it will be found to be covered with very fine and regular lines running from the beak to the opposite edges. So fine and close-set are these, that a line an inch long, drawn across them at right angles, will cross about ninety of them. The valves are more or less covered with a colouring of rusty red, relieved by numerous long sharp wedges of white. These are on parts or the whole of the shell, sometimes so plentiful that there is no room for solid masses of the red colour, and it only shows in zigzag lines.

It is difficult to get two shells that at all agree in the distribution of white and red, and even the two valves of the same shell will differ widely in this respect. The interior, also, exhibits characters sufficiently striking to prevent its misidentification. There is a broad flange below the hinge, whereon are cut about twenty teeth, in two series. Immediately below the ligament is a smooth space, clear of teeth, but these are arranged in a row of about ten on each side of this space. As the shell grows the flange lengthens, and more teeth are added to the ends of the rows farthest from the beak; but those nearest the smooth central space are being rubbed down or absorbed at the same rate, so that the net increase is about *nil*. The free edge of the valves, internally, has a series of raised marks, like the tips of the teeth of a comb, and it is from these the creature gets its name. The pallial impression is much deeper than those of the muscles at each end of it, and it is uninterrupted by any sinus. This shell is about two inches in length.

We must not omit to mention a group of shells that are fairly common upon many shores, and are usually found among the bucketful the children have collected.

First of these is the bold Rayed Trough-shell (*Mactra stultorum*), more plentiful in the north than the south of Britain. The various species of Mactra are inhabitants of sand in deep-water, but their shells are freely cast up on the shore. These are smooth, except that the annual periods of rest from shell-making is plainly marked in deep concentric grooves. Like that of the Spiny-Cockle, the foot of Mactra can be ex-

RAYED TROUGH-SHELL.

tended and used like a finger, and also as a leaping-pole. They are destroyed in great numbers by star-fish, and many empty valves may be found with the clean round boring that shows the animal fell a victim to the whelk. *M. stultorum*

is usually coloured some shade of brown, with a number of
white bands radiating from the beak.   The hinge arrangements in this genus are worthy of note, as indeed they
are in all the genera, and must be carefully studied by anybody
who wishes to have anything more than the merest superficial
knowledge of conchology.

In the Trough-shells the ligament of the hinge is short and
thick, and contained in a spoon-like cavity in each valve.
Immediately in front of it there are two shelly teeth, joined
above in the form of a Λ, and from each side of the beak there
runs off a strong ridge-like tooth, the surface of which is
"milled" like the edge of a shilling or a sovereign.   The
Elliptic Trough-shell (*M. elliptica*) is not so triangular as *M.
stultorum*, and is without the white rays.   The Cut Trough-
shell (*M. truncata*) might be appropriately styled the hatchet
shell, for its truncated ends give it a very close likeness to the
head of a hatchet.

Related to the Trough-shells are the Otter-shells (*Lutraria*),
of which we have two species.   They burrow in the mud and
sands, of estuaries especially, and are found from low-water
to about twelve fathoms.   Having found a complete, though
empty-shell, you will be surprised to discover that it will not
close properly, and you not unnaturally suppose that you have
got hold of a malformed specimen, whose shell has got a twist
somehow.   That, however, would be a mistake, as you would
find when other specimens came in your way, and you found
they all had the same objection to closing at the ends.   From
one end, when the creature is alive, protrude its united
syphons, large and thick; and from the other end is thrust
out the useful "foot," with which its burrowing is effected.
Where you happen to find the usually broken valves of the
Otter-shells, it is worth while to dig in the muddiest spots
thereabout at extreme low-water, and you will probably be
rewarded with perfect specimens, and have the greater satisfaction of seeing the living creature within.

Then there are the Tellen-shells (*Tellina*), a bright and delicate-looking group, with shells that appear as though they had been subjected to considerable pressure. They are finely grooved with concentric lines, and decorated with broad bands of pink. One of the most plentiful of these is the Thick Tellen (*Tellina crassa*), in which the pink bands radiate from the beak across the shell. Thick is a comparative term, and is so used here, for the shell, as compared with a *Mactra*, for instance, would be considered rather thin; but in contrast with other Tellens, it is solid and substantial. The interior is delicately tinted with pink or orange. The pallial sinus is large and rounded. The Fragile Tellen (*T. tenuis*) has thin shells that are very easily broken. Its surface is very smooth, of an orange tint marked with bands of pink and white. There are half-a-dozen other British species. The Tellens burrow slightly in sandy mud, frequently in shallow water. They may be dug for on a suitable beach between tide-marks, though their range extends to about fifty fathoms.

Somewhat similar to the Tellens in their delicacy and style of ornament are the Sunset-shells (*Psammobia*), so called on account of the crimson patch around the beak, from which rays of a similar hue run off to the margin. If the shell is so placed before you that the beak is downwards, these rays suggest the far-reaching rays from the sun that streak all the western sky, when Sol dips below the horizon for the night. There are four British species. The two ends of the shell are nearly equally rounded, but in an allied genus—

The Wedge-shells (*Donax*), the hinder end is much more acute than the front, so that their popular name is very suitable. They have a suggestion of sunset rays, too, but not so strong or so symmetrical as in *Psammobia*. The most familiar species is the Common Wedge-shell (*Donax anatinus*), which may easily be distinguished from the others by the milling of the interior edge of the valves. The Polished Wedge (*D. politus*) may be equally well separated by its superior gloss, and by a white band which runs backward from the beak.

Then there are the familiar Razor-shells (*Solen*) that must be dug out of the sand at low-water; and quick work you will find it, if you succeed in catching any specimens. Very good examples may often be picked up on a wide sandy beach, but minus the animal. They are sought for food, and the professional catchers are very expert in their movements—they need to be, or the business would not pay a dividend. Everybody knows the razor-handle-like shells of *Solen siliqua*, if they have no acquaintance with the animal. They are flattened cylinders, widely open at each end for the extrusion of the foot and the syphons. The hinge is near the front extremity of the shell, the ligament in a full-grown specimen measuring an inch and a half. There are two teeth in each valve, though some have three in the left; but it is difficult to pick up empty shells in which the teeth are intact. The Razors spend all their lives buried vertically in the sand. When the sands are covered by water they rise to the mouth of their burrow and protrude the syphons, but those that are situated so far in shore as to be uncovered at low-water, then plunge in to a depth of a foot or two. They never leave their burrows, except on compulsion, in the shape of the salt and spade of their enemy, the fisherman. The species, with a very straight margin to its shell, is the Pod Razor (*S. siliqua*) which attains a length of eight inches; that with a distinctly curved outline is the Sabre Razor (*S. ensis*).

A brief glance at some borers and excavators must suffice to close this long chapter. The small, upper figure in accompanying plate is the Red-nosed Borer (*Saxicava rugosa*), a species that largely helps the sea in its ceaseless attacks upon the coast line. It is the office of the Borer to excavate cells in the face of the rock, and as it is never solitary in its work, but attacks a rock in "gangs," as a human excavator would put it, the result is the complete honey-combing of the surface. They may often be found free in crevices of the rocks and about the roots of seaweeds—that Alsatia for a very varied

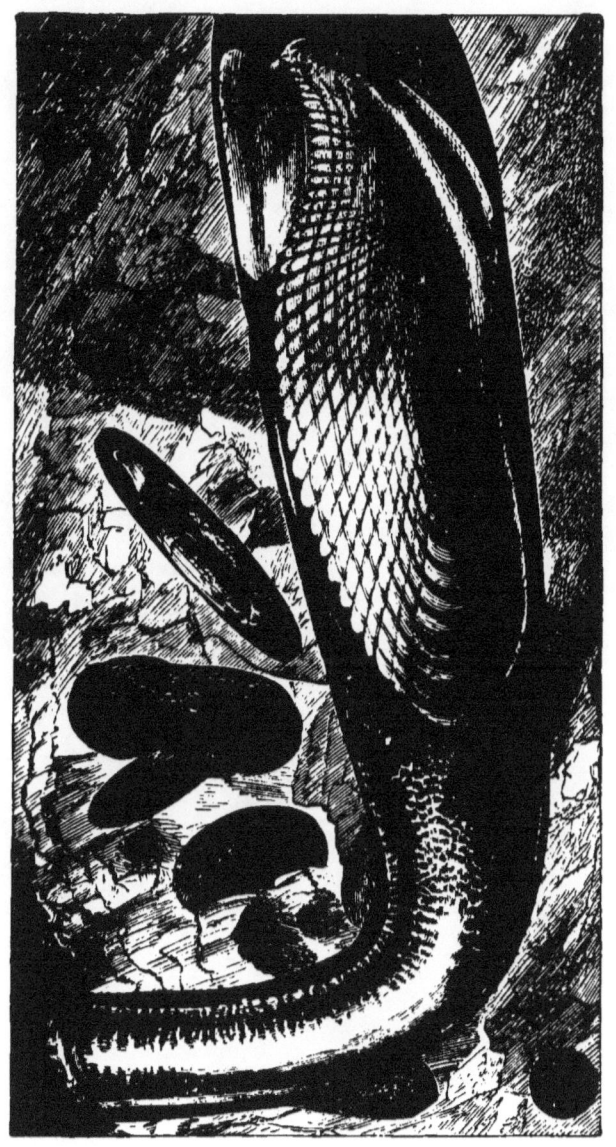

RED-NOSED BORER.

PIDDOCK.

population. The shell has a distorted look about it, and the
valves will not fit properly, the ends gaping to allow the foot
and the syphons free play. It is very variable, however, and
consequently has been a splendid subject for the variety-
mongers and species-splitters, who have manufactured quite a
long list of species and genera out of it. It changes a good
deal at different periods of its life, and thus affords opportunities for careful descriptions made from isolated specimens
utterly disagreeing with each other; therefore, the individuals
described must belong to different species! In its early state
the shell is symmetrical, and has two minute teeth in each
valve; but before it has reached maturity it has lost its claim
to be considered graceful or symmetrical, and has either worn
its milk-teeth out or abandoned them as useless. The shell is
covered with ridges and wrinkles, and it is by their help that
it carves out its chamber in the rock. Sometimes on turning
aside a curtain of weeds from a rock-face you will see a large
number of crimson points, which, however, instantly disappear
if they have been ever so lightly touched by the weeds. These
are the ends of the borers' syphons, protruded from their
ventilation holes; they are united almost to their extremities,
and present the appearance shown in our figure.

The Piddock, or Finger Pholas (*Pholas dactylus*), is a much
larger species with some difference of structure. Its pure
white shell, though thin and fragile, is covered in front with
rasp-like ridges, and by means of it the chambers and tunnels
of the rock are bored. Holding to the rock with the clear
crystalline foot, the Pholas gives its shell a swing half-way
round in one direction, then a swing back, and so by alternate
half-revolutions, the rasping of the shell gradually excavates
a chamber sufficiently large to shelter it, communication with
the outer world being maintained by the large syphons. So
far there is no very great difference between the Pholas and
the Saxicava; but the Pholas is peculiar in that it possesses
neither ligament nor hinge, and in addition to the orthodox

two valves, it has some additional ones. The hinge-plates are reflected back over the beaks, and a powerful muscle is attached thereto to keep the ordinary valves together. Above these, and to protect this muscle, are two short accessory valves, and a third, which is long, and extends back over the dorsal edges of the big valves. In other species of Pholas these arrangements give scope for variation.

And now it is time we gave some thought to the one-valved and valveless mollusks of the shore.

CHAPTER XIV.

SEA-SNAILS AND SEA-SLUGS.

MOLLUSKS that have their shell all in one piece are technic-ally known as the Gasteropoda, or belly-footed creatures; but for our purpose the term sea-snail will serve admirably, for it is a popular term that· will not cause misunderstandings, as many popular general terms do. The sea-snails, as living creatures, are more amenable to study by the shore-naturalist, than is the case with the bivalves; and every rock, whether it be thickly clothed with weeds, or bare and exposed to the full fury of the waves, will provide us with specimens. It is true, that all visitors to the sea-shore are well acquainted with the most plentiful of these—the periwinkle, the purple, and the limpet. But though they are familiar with the forms and names of such common objects, there may be among my readers some to whom the principal facts in the economy and structure of these species may be new or interesting.

I fear, that in popular estimation, there is but one kind of Limpet. As a matter of greater exactitude, I may say that eight or nine species may be found on our shores; and we may find some points of interest even in the too common species (*Patella vulgata*). Only those perhaps who have been badly in want of bait for a little fishing have troubled to see what is beneath the conical shell; but the shell itself is worthy of a little attention. What could be better adapted for the animal's mode of life than this? The Limpet is not a deep-water mol-lusk, but lives between tide-marks, where it receives the full force of the waves as they beat and hammer the rocks in

o

stormy weather. But the Limpet has a broad foot, which
exudes a thick glue, whereby it sticks tightly to the rock.
Then his muscles are powerful, and by their aid he pulls his
conical roof well down till its edges fit into the little pit he has
sunk in the rock surface, and thus ensconced he can defy the
hardest gale that may chance to blow and the heaviest water-
hammers that the sea uses against the land. The Limpet is

LIMPETS.   PURPLES.

typical of the Briton, alike in his tenacity of purpose and his
love of privacy. But with all his exclusiveness John Bull likes
to open his doors and windows wide to let in the air, and we
shall find the Limpet resembling him in this detail; for if you
seek him when the tide is out, you may surprise him with his
roof so lifted up that the edges are a quarter of an inch away

from the rock. Then is the time to take him unawares, and force his foot from its firm hold. Having secured him, we are at liberty to inspect the owner of this strange house, but we can best do this by placing him in our clear glass bottle, and letting him crawl up the side.

That which is known as the mollusk's "foot" has no relationship with the feet of vertebrate animals, the name being suggested by the similar use to which dissimilar organs are put. We have already explained that the term gasteropod signifies "belly-fcot," and if we were to cut through the "foot" of the Limpet, we should find that it is indeed its belly, for it contains the principal portion of its viscera. We are not going into the anatomy of the mollusca, just now, but will confine our attention to its exterior. It has now begun to climb up the glass, and we can see that the foot is spread out so that it occupies the greater portion of the area covered by the shell. At the fore part it has a distinct head, with a pair of tentacles, ditto eyes, and a very evident mouth, for the Limpet's principal occupation appears to be to lick the surface upon which it is gliding. Around the foot and the head there runs a frill which is really the creature's breathing apparatus, and between that and the shell there is, of course, the mantle by which the shell was secreted, and is enlarged as occasion requires. The Limpet is now in rapid motion, and we can see that it progresses in exactly the same fashion as do the garden snails and slugs, that is, by a series of muscular contractions, evidenced by the constant ripple along the surface of the foot. The foot exudes a very tenacious slime, which enables it to obtain perfect contact with the surface over which it is gliding, or upon which it is resting. It is perfectly astonishing how much nonsense is still written in books upon this subject by persons who ought to know better, and who could easily test the correctness of their views by occasionally studying Nature, instead of relying so much upon academical teaching, and that of an antique character. Their statement is, that the

Limpet holds on so tightly by creating a vacuum, some say under the foot, others under the shell. So ancient an authority as Reaumur disproved these notions. He tested the matter by cutting a Limpet in two, shell and all. According to the teaching of the vacuumites, the animal's hold should then have loosened; but no, the two portions still adhered to their base. Anyone by observation can testify to the truth of Reaumur's explanation; there is the same powerful hold in the foot of a garden snail on a damp surface, but in that case it does not seem so great, because his shell affords a better hold for the experimenter. The annoying feature of the Limpet is the shape of his shell, which prevents our taking hold of it. Where the surface of the rock is friable, as some of our Cornish Killas rocks, and the chalk rocks of the Kentish coast, the Limpet's foot, when forcibly pulled up, brings with it particles of the surface, which have separated from the parent rock more easily than from the glue of the mollusk's foot.

A wonderful thing about the Limpet is its power to sink a shallow pit in the surface of the rock, corresponding to the shape of the shell; and this, of course, has led to much theorising to explain how it is accomplished. Patent solvents secreted by the animal, the carbonic acid gas given off from the breathing apparatus (which strangely does not destroy its own shell!), and so on. A little study of Nature would show that the wonderful organ which enables them to scrape away the surface in long zigzag lines, as they crop the minute vegetation, would be equally effective if applied to the spot upon which they prefer to roost, and to which they habitually return after their pastoral wanderings. The action of this tongue on the rocks can be very distinctly *heard* on the shore, though possibly not in the library or the museum, where only the empty shells are admitted. It is worth while dissecting a Limpet, and getting out this remarkable tongue, which is a ribbon-shaped organ, closely studded with minute hooks of flint, to the number of nearly 2,000. A similar *lingual ribbon*, as it is termed, will be found in most of the Gastercpods.

I have dealt at such length with the Limpet, because its structure will enable us to understand the other mollusks we have to mention, widely as they may appear to differ in the forms of their bodies and shells. The Limpet's shell is a low cone, and the shell of a Whelk is a greatly elongated cone, coiled spirally upon itself; the animal adapting itself to that form.

In addition to the Common Limpet (*Patella vulgata*) we have the Smooth Limpet (*Patella pellucida*), which must be sought at low-water on the borders of the laminarian zone. It feeds upon the Great Oar-weed, and a peculiar variation will be found between the specimens feeding on the smooth flat fronds and those feeding on the great stems. The shell of the first is coloured a pale brown, pellucid as its specific name suggests, the apex set very far forwards, and from it there start backwards from three to six exceedingly fine radiating lines of a dazzling brilliant blue. The specimens that live upon the Oar-weed's stems look entirely different, for the shell becomes

SMOOTH LIMPET.   SMOOTH LIMPET, THICK VARIETY.

thickened, and consequently much more opaque, and its shape alters to enable it to sit close on a rounded surface. It was formerly considered a distinct species, and was named *Patella lævis*. So, too, the little Tortoise-shell Limpet (*Acmæa testudinalis*), changes its form when feeding upon the leaves of the Grass-wrack (*Zostera marina*), and has then had the name of *Acmæa aivea* bestowed upon it.

There are other forms of Limpets (though not species of Patellidæ) to which we wish to refer, but we are getting far away from our illustration of the Purple (*Purpura lapillus*), on page 208, to which we must now hark back. The Purple is often known as the Dog Winkle. It abounds upon the rocks between tide-marks, whence it may be picked without the formalities necessary in the case of the Limpet. It comes off easily, for its foot is small, but the moment it is disengaged from the rock it retires into its shell and closes its door. Now apart from the difference in the shape of the shell, here is another departure from molluskan arrangements as illustrated by the Limpet. It is called an *operculum* (Latin, a cover or stopper), and is so attached to the foot, that when the Purple withdraws from public view this comes last, and fits the mouth of the shell so accurately that there is no getting inside. In this case it is a horny oval disk, but in some species it is strengthened by the deposit of layers of shelly matter until it becomes of considerable thickness and quite stony. If we mark our disapproval of the Purple's lack of courtesy in slamming his door in our face, by pushing against his door, he retaliates by exuding a purple fluid, which is said to permanently dye fabrics a similar hue. The Purple is not a vegetarian like the Limpet. His mouth forms a fleshy proboscis, which contains a marvellous boring apparatus— the modified tongue. Often you may pick up bivalve shells on the beach, of which one has been pierced with a very clean and smooth round hole near the beak. If you did not know otherwise, you might suppose that this was the work of a person who desired to make a shell-necklace or other orna- ment, and had bored this hole with the greatest of care, and then had unfortunately dropped it on the beach. The truth is, it is the work of the Purple, or some other carnivorous sea- snail. He has the reputation of being very destructive to mussel-beds, by boring these workmanlike holes in their shells, and literally eating the poor mussel out of house and home. That is the style in which the Purple gets his living; but he has a Nemesis in the shape of the Star-fish, and I have seen one Star-fish eating or digesting three Purples at once. It is a case of "diamond cut diamond," for you would think a Mussel or a Limpet would be safe enough with the shell closed down, and so you might suppose the Purple's operculum would shield him from the Star-fish; but as I have already described in an earlier chapter, the Star-fish knows well how to deal with obstinate victims who won't show their noses outside the door when their enemy calls—he digests them first, and swallows them afterwards. Here is a complete reversal of the Shakespearean motto, " May good digestion wait on appetite;" to be complimentary to the Star-fish we should say, " May appetite on good digestion wait!" In the bottom left-hand corner of the purple-and-limpet illustration, is a baker's dozen of nine-pins: they are the egg-cases of the Purple, which may be found in larger or smaller patches on any rock where these mollusks abound.

Among the weeds on the rocks we are sure to find the Netted Dog-whelk (*Nassa reticulata*), with a rather dirty-looking shell. It is covered with broad grooves crossed by fine lines at right angles, producing the appearance of network, which gives it the distinctive name, netted. Its scientific name also is suggested by the same

NETTED DOG-WHELK.

appearance, for *Nassa* is Latin for a special kind of fishing-net. Like the Purple, the Dog-whelk is carnivorous. There is a prettier species, with a thick lip, called *Nassa incrassata*.

The true Whelk (*Buccinum undatum*) only comes within our province in the shape of empty shells cast up on the beach, for its range is from low-water to a hundred fathoms. In deep water it is very plentiful, and fishermen who want it for bait, let down baskets containing pieces of fish, which attract a large number to their doom. Their remarkable clusters of egg-nests are frequently washed ashore with seaweeds; each capsule in the bunch contains about half-a-dozen eggs. The shell of the Whelk, rubbed down on a smooth slab of stone, affords an admirable vertical section illustrating the structure of gasteropods.

The Red Whelk is the *Fusus antiquus,* so-called because it abounds in a fossil condition in the Red Crag of Essex, where also occurs a reversed form—that is, with the spire coiled the contrary way, and hence called *Fusus contrarius.* In Scotland

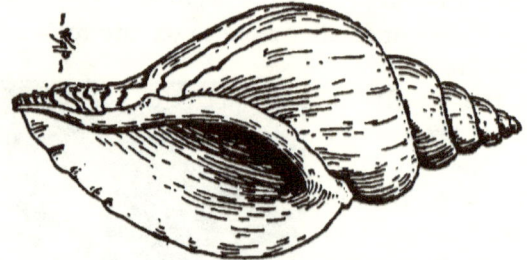

RED WHELK.

it is the Buckie, or the Roaring Buckie, for this is the shell in which the roar of the sea resides. It is more esteemed than the Common Whelk as food by the poorer population of Scotland. It occurs, like *Buccinum,* from low-water to a hundred fathoms.

There is a fairly common shell, similar in size and general form to the Purple, but bristling all over with flattened re-curved hooks, in clusters of threes. It is generally known as the Sting-winkle (*Murex erinaceus*), one of a genus from which the celebrated purple dye of ancient Tyre was obtained. Its familiar name it owes to its sharing in the hideous crime of destroying edible species for the sole purpose of gratifying its own base appetite. The fishermen have actually noticed it in the act, and seeing the peculiar boring apparatus at work, have thought this a sting. It is far worse than that, for a sting may be survived, but no mollusk, I believe, gets over the attack of the boring tongue, which changes its function when the boring is finished, and becomes an instrument for tearing and masticating its victim's flesh.

The exotic representatives of the great Cone-family of shells are familiar and admired objects in collections as well as on nick-nack tables in the drawing-room. We have no native species of the genus *Conus*, but we have a number of representatives of the *family* in the Pleurotomas and Mangelias, though they do not approach very closely to the typical form of a Cone-shell, with which we commonly associate the spotted Cone (*Conus marmoreus*) of Chinese seas. The Seven-ribbed Conelet (*Mangelia septangularis*) is like a tiny Buckie-shell —it is but half an inch long—with bold longitudinal ribs, of which you can count seven in one revolution of the shell. The shell is thick, of a dull pinkish hue, and unprovided with an operculum. The outer lip is notched where it joins the previous whorl. There are several British species.

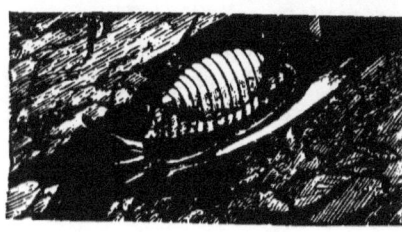

LOWRY.

One of the most charming of our native shells is the little Cowry (*Cypræa europea*), which is so plentiful on most of our shores. Most of us who have visited the sea-shore in childhood have had the delight of hunting for this shell, empty and clean, among the ingredients of a fine beach; but probably some of those who are most familiar with it as an empty shell would scarcely recognise it for the same species if they saw the living Cowry gliding along with his shell on his back. He carries a pair of tentacles, with eyes at their base, and the long curved tubular tongue ready for service; but the most singular feature is that his mantle is used not merely to clothe the delicate body, but a portion of it comes outside, and closely wraps the greater part of the shell. In its younger days it had not its present beautifully arched lip, which almost closes up the doorway of the shell, and leaves but a narrow slit, delicately denticulated, to allow the foot and mantle to pass through. Before maturity it had a wide mouth, with a sharp thin edge to the outer lip, but that, you see, has now grown over towards the inner lip. The colour of the shell may be described as a flesh-tint on the upper surface, varying in intensity to both lighter and darker. Many specimens bear on the crown of the shell three ill-defined blotches of a very dark brown. The under surface is white. The whole shell is ornamented by very regularly disposed transverse ribs, which are rounded and polished.

There are several other cowry-like shells to be found generally distributed, but by no means so plentifully on our coasts. One of these is the Smooth Margin-shell (*Erato lævis*), smaller than the Cowry, and with the lip curved outward, instead of inward as in the Cowry: it has thus an external margin, whence the name. It is white and exquisitely smooth. The animal is very similar to *Cypræa*, and it envelops its shell in the same fashion. Of similar habit is the Poached Egg (*Ovula patula*), though the shell is very different. The mouth of the shell gapes widely, and the lip is thin and sharp. Its colour is white with a pink tinge, and its appearance is so suggestive of its name that there is little likelihood of mis-identification. It is a South Coast form.

A solid-looking shell, with a highly-polished surface, over which three lines of arrow-heads are chasing each other, a perforation of the shell just outside the inner lip, a fairly wide mouth, closed when at rest by an operculum: these are the principal features of the Necklace Natica, so-called because it deposits a large number of eggs, so agglutinated into a broad spiral band, that the whole has been likened to a necklace. So it is called *Natica monilifera*, and *monilifera* means neck-lace-bearing. The animal is an odd creature, whose mantle laps partly over the shell, and the large foot is furnished in front with a broad fold, which is turned back as a protection to the head. It is herbivorous, and crops the seaweeds on sandy and gravelly shores, from low-water to about ninety fathoms.

There is a very thin, ear-shaped shell, clear and fragile, known as *Lamellaria perspicua*. It is not sufficiently capacious to accommodate the whole of the animal, so parts of it have to remain permanently outside; the mantle, for instance, can-not be withdrawn, and it folds over, completely wrapping up the shell and hiding it from view. It is an awkward thing to have your house so small that you cannot get right inside, because in the sea there are so many hungry creatures always roving about, and snapping up any delicate morsel that is unprotected; and even some that are protected get swallowed up in like manner. But *Lamellaria* has learned how to make up to some extent for Nature's stinginess in the matter of shell-stuff. About a quarter of a century since, Giard showed that *Lamellaria* was to be found in association with com-pound ascidians, a group to which we shall call attention in a later chapter. Quite recently* Prof. W. A. Herdman, Director of the Port Erin Biological Station, added greatly to the interest of Giard's observation by one of his own, which shall be given in his own words:—

"*Lamellaria perspicua* is not uncommon round the south end of the Isle of Man, and is frequently found under the circumstances described by Giard; but I met lately with such a marked case on the shore near the Biological Station at Port Erin, that it seems worthy of being placed on record. The mollusc was on a colony of *Leptoclinum maculatum*, in which it had eaten a large hole. It lay in this cavity so as to

---

* "Conchologist," 1893.

be flush with the general surface; and its dorsal integument was not only whitish with small darker marks which exactly reproduced the appearance of the *Leptoclinum* surface with the ascidiozooids scattered over it, but there were also two larger elliptical clear marks which looked like the large common cloacal apertures of the Ascidian colony.   I did not notice the *Lamellaria* until I had accidentally partly dislodged it in detaching the *Leptoclinum* from a stone.   I then pointed it out to a couple of naturalists who were with me, and we were all much struck with the difficulty in detecting it when *in situ* on the Ascidian.

HORN-
SHELL.

"This is clearly a good case of protective colouring. Presumably the *Lamellaria* escapes the observation of its enemies through being mistaken for a part of the *Leptoclinum* colony; and the *Leptoclinum* being crowded like a sponge with minute sharp-pointed spicules is, I suppose, avoided as inedible (if not actually noxious through some peculiar smell or taste) by carnivorous animals which might devour such things as the soft unprotected mollusc. But the presence of the spicules evidently does not protect the *Leptoclinum* from *Lamellaria*, so that we have, if the above interpretation is correct, the curious result that the *Lamellaria* profits by a protective characteristic of the *Leptoclinum* for which it has itself no respect, or to put it another way, the *Leptoclinum* is protected against enemies to some extent for the benefit of the *Lamellaria* which preys upon its vitals."

Since the publication of Prof. Herdman's note, I have frequently found *Lamellaria* on the undersides of large stones at low-water on the Cornish coast.   I have always found it on *Leptoclinum gelatinosum*, and can quite endorse his remark as to the difficulty of distinguishing it.   On one occasion I found no less than four specimens feeding upon one patch of the ascidians, and pointed them out to a friend, who, however, failed to see them until they were absolutely touched by my finger. The shell is exceedingly delicate, and in the hands of most persons would be hopelessly ruined at the first touch. The ordinary methods adopted by conchologists for getting the animal from the shell will not answer in this case; but I have a plan which succeeds admirably. I give a specimen of *Lamellaria* to an anemone of refined tastes, who will deal with it carefully. *Bunodes verrucosa* is my favourite assistant, and he returns the shell clean and sound in a day or two.

There are several species of Spire-shells (*Rissoa*) to be found feeding in great numbers on Grass-wrack and Sea-lettuce, and we shall also find the empty shells in the sand. There are, however, other forms that may be confused with them and with each other, that are very plentiful in sand. These are the comparatively large Turret-shell (*Turritella communis*), which is ornamented with spiral ridges, each one running continuously from the apex to the mouth. In the Ruddy Pyramid (*Chemnitzia rufescens*), which is much smaller, but of similar form, the ridges run *across* instead of *along* the whorls, whilst in the Horn-shell (*Cerithium reticulatum*), a similar effect is obtained by several rows of very regularly arranged round dots in high relief. A more distinct member of the family of Cerites is to be found in the well-known Pelican's-foot or Spout-shell (*Aporrhais pes-pelicani*), in which the whorls are boldly tuberculated When the shell has grown to its full length, its annual stages of growth take a somewhat different direction, and spread out in expansive lobes and corrugations until it bears a fanciful resemblance in outline to the foot of the pelican. The shell is about an inch and a half in length, and very thick. The animal is carnivorous.

Delicate specimens of the well-known Wentletraps (*Scalaria*) may be found among fine sands. They are readily known by their dazzling whiteness, the nearly round and flat-lipped mouth, and the bold curved ridges that stand out across the whorls like cogs on a wheel. To this genus belongs the

PELICAN'S-FOOT.

Precious Wentletrap (*S. pretiosa*), from China, for a single specimen of which as much as forty guineas has been paid. This was in the days when shell-collecting without any scientific object in view was a mania with some wealthy people; just as we have had the tulip-mania, and now have the orchid-mania affecting persons who are impelled by fashion rather than a love of knowledge or the beautiful in Nature.

However we may be inclined to pass over the Periwinkle (*Littorina littorea*) as a species too common to need any attention, it is bound to thrust itself upon our vision at every turn among the rocks, where it swarms. It appears strange that whilst this species is so largely eaten by the poorer classes in towns as a "relish" for tea, the allied and almost equally common species, *L. rudis*, should be let severely alone. But the explanation is probably to be found in the fact that whereas *littorea* deposits her eggs in the ordinary way, *rudis* retains hers until they have hatched out. Now seeing that the Winkles of both species develop their hard stony shells before they hatch, it would be impossible to eat *L. rudis* without the great inconvenience of having these hard gritty infants damaging one's teeth. The smaller red, or bright yellow shell, that may be found in abundance on the rocks and weeds between tide-marks, is *Littorina littoralis*.

The seeker for shells on a sandy shore must do as the children do—throw himself prone upon the beach, and hunt thoroughly, inch by inch, examining the topmost layer first, then lightly scraping it off and bringing fresh treasures to light. In this manner he will certainly turn up the exquisite little

Pheasant shells (*Phasianella pullas*), that have the misfortune to be so small, or they would be greatly esteemed for their rich colouring. They are very smooth, and of a white or pale yellow hue, but so thickly covered with fine crimson lines that at first sight this appears to be the colour of the shell. These lines run parallel with each other, but with many curves, some flowing gently, others short and acute. These lines vary much in thickness throughout their length, here being very fine, there thickening gradually and thinning off again. The shells are less than a quarter of an inch in length, and the mouth is closed with an operculum. The animal has the peculiar habit of moving first one half, then the other, of its foot in progressing.

One of the handsomest of our common rock-shells is the so-called Common Top (*Trochus zizyphinus*), though it is scarcely as plentiful as the much smaller Grey Top (*Trochus cinereus*).

THE COMMON TOP.

It is pyramidal in form, with an almost flat, broad base; the mouth closed by a spiral, horny operculum. In some species there is an umbilicus, in others it is wanting. The animal has two small fringed lobes between the tentacles, and similarly fringed lappets to the neck. The sides, too, are lobed, and several tentacle-like processes are given off from them.

The Grey Top (*T. cinereus*) is variable in colour. Usually
it is a dull yellowish grey, with inconspicuous dark zigzag
marks upon it; sometimes the ground colour is pinkish-white,
with decided pink markings, which present a checkered
appearance. There is a deep and wide umbilicus. In *Trochus
zizyphinus* there is no umbilicus, and in the large Painted
Top (*T. magus*), again, there is a very wide one. This last-
mentioned species lacks the smoothness of outline exhibited by
the other two, its whorls being more boldly ridged at their
junctions (*suture*). The animal has the head-lobes largely
developed, and it is brilliantly and variously coloured; hence
its name. The Tops are vegetable feeders.

On our South-western shores, when strong winds have blown
from the S.W. for days together, there are borne to us on the
waves, and wrecked upon our beaches, singular sea-snails from
the mid-Atlantic. There the Violet-shells (*Ianthina*) float in
myriads, and consume the still more plentiful "Sallee-man"
(*Velella*), a Jelly-fish we have mentioned in a previous chapter.
There are many singular features about this *Ianthina*. Like
a shipwrecked mariner, it constructs a raft, secreting glutinous
material from the foot, in the form of many air-chambers

VIOLET-SHELL.

RAFT OF VIOLET-SHELL.

cemented together, and bearing beneath a large number of
egg-capsules. The shell is of somewhat similar shape to that
of the Tops, but with a much larger mouth. Its material, too,
is so thin it can almost be seen through; and on the upper
part it is white, whilst beneath it is coloured violet, whence
its names. The animal has its head produced into a thick
muzzle, with a pair of tentacles and a pair of eye-stalks,
but no eyes. The breathing organs are two plume-like gills
which protrude from the shell.

We must return now to certain limpet-like forms, of which one, the Key-hole Limpet (*Fissurella græca*), might be easily passed by as a Common Limpet that has got damaged. In form and appearance the shell is not unlike the common kind; the peculiarity consists in a short and narrow slit at the summit, which has suggested the name. As a living mollusk it must be sought in the laminarian zone, but the empty shells are to be found between tide-marks. A smaller, but not very dissimilar shell, has the key-hole not on the apex, but a little in advance of it. This is the Perforated Limpet (*Puncturella noachina*). It is rarer than the last, and is to be sought on the North coasts, where it lives below the twenty-fathom line.

Yet another species is depicted in this cut. It is the Slit Limpet (*Emarginula reticulata*), in which the notch or slit is in the fore edge of the white shell; length of slit variable. Internally the shell is thickened near the notch, and outside it is deeply grooved, so that strong ribs radiate from the summit, and are themselves partly cut up by lighter grooves transversely to the others. It comes up to low-water mark, so may be taken alive from the shore. There is a second British species, the Rosy-slit Limpet (*E. rosea*), much smaller, and

SLIT LIMPET.

sometimes with the slit rosy, but this is not a reliable character.

On our Southern shores may be found—frequently on oysters —a shell that may be said to be the highest development of the limpet type. Seen from the point of view taken for our illustration, there is a long gentle curve from the mouth to the beak, which is spirally twisted. The general effect is to remind one of the conventional representations of the Cap of Liberty. Owing to this cap-like form, it is known as the Hungarian Cap, and the Torbay Bonnet (*Pileopsis hungaricus*). In colour it varies from brown to an indefinite dirty-white hue.

The miniature elephant's tusks represented in the illustration with the Hungarian Cap are really shells, called Tusk or Tooth-shells (*Dentalium*). They are represented of the natural size. The shell is open at each end, and is tenanted by a strange little animal who is at-tached to it near the small end. The Dentalium is not a very highly devel-oped creature, for though it has a head, it is quite a rudimentary one, without eyes. But though it lacks eyes, it has a mouth, surrounded by eight tentacles, and into this go foraminifera and other minute crea-tures it picks up on the sands and mud in deep water. We have two British species, of which

HUNGARIAN CAP.   TUSK-SHELL.

we may occasionally find the empty shells washed up on the sands. Of these the Elephant's-tusk (*Dentalium entalis*) is very smooth and quite white throughout; whereas the Grooved-tusk (*D. tarentinum*) is delicately grooved at the larger or fore end, and tinged with pink at the small end.

In chipping off fragments of rock at low-water, upon which anemones and other specimens are sitting, you may often get more than you had thought, for sometimes when the piece of rock is placed in an aquarium, other creatures will make their appearance, which were unobserved before, owing to their colour, and the closeness with which they attach themselves. One of these is the Bristly Mail-shell (*Chiton fascicularis*), distinguished from other British species by the possession of little bunches of short bristles, which are arranged along the shell-border opposite each plate of mail. There is considerable resemblance between these creatures and limpets, though there are also important differences. Instead of the shell being in one piece, it is composed of eight transverse plates, which overlap at their edges, and allow it to be rolled up like a woodlouse. Each plate is attached to the mantle by its front margin, and the mantle forms a narrow border all round the shell. The animal, like the limpet, has a broad foot upon which it creeps, mostly at night, so far as my observations of *C. fascicularis* go. Its head chiefly consists of its mouth and jaws, eyes and tentacles being dispensed with as unnecessary

SMOOTH MAIL SHELL.

to its manner of life. The breathing organs are similar to those of the limpet, but are arranged round the posterior end of the body only. The shell is very flexible in all directions, so that the animal is not constrained, like the limpet, to return to the same roosting spot each time it wishes to rest.

There are a number of British species; the one figured is known as the Smooth Mail-shell (*C. lævis*). It has a glossy shell of a reddish hue, with a central ridge. The largest of the native forms is the Marbled Mail-shell (*C. marmoreus*), whose delicately sculptured shell is further ornamented with a mottling of browns and yellows. It is about an inch and a quarter in length. The British species is almost as long, but of much more slender proportions. The most plentiful form is the Grey Mail-shell (*C. cinereus*), which does not greatly exceed half an inch in length. It is not entirely grey, though this is the prevailing tint, but there are delicate mottlings and streaks of many colours upon it.

We now reach what we may very fitly term the Sea-slugs, for they are creatures that externally have considerable resemblance to the land-slugs, though structurally they are very different, and they are far removed from each other in classification. The land-slugs (*Limax*) carry a little shell embedded in their back, and their breathing organs are internal; the Sea-slugs are entirely shell-less, except in the embryo-stage, and their breathing apparatus is always exposed, and situated on the back or sides. In consequence of this characteristic, the Sea-slugs, as a group or section of the Gasteropods, are called the Nudibranchiata, or naked-gilled mollusca. They are plentiful on rocky coasts, where they range from half-tide to

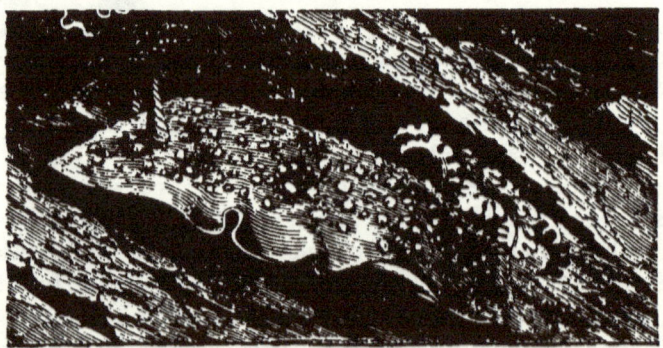

SEA LEMON.

a great depth. The best plan is to seek for them at low spring-tides, turning over stones at the edge of the laminarian zone, when the slugs will be found at rest on the under surfaces, in a more or less collapsed condition. They will readily respond, however, to the attention paid them by putting them in the calm clear water of our collecting bottles, and extending their tentacles and branchial plumes, will explore their new quarters. One of the most striking of these sea-slugs is the Sea Lemon (*Doris tuberculata*), which is about three inches in length, broad, and with the upper surface thickly studded with tubercules; this, in conjunction with its colour, gives it a very close likeness to the half of a lemon adhering to the rocks. As will be seen in the illustration, there are two tentacles, and these are retractile within special cavities. The branchial plumes are arranged in a crown-like circle in the middle of the back, but near to the posterior end; and these also can be withdrawn at the creature's will. There are several British species, some of them very small, and they range from low-water to twenty-five fathoms, feeding upon zoophytes, sponges, anemones, and their own kind.

Doris johnstoni is a smaller species than *tuberculata*, but is worthy of attention on account of a certain resemblance to *Lamellaria*. It is "got up" to mimic a sponge. There are no tubercles on its surface, which is very finely roughed, so that it is sponge-like to the touch. In colour it is creamy, wonderfully speckled with larger and smaller spots of pale brown, that produce the effect of the porous surface of a sponge, and the large spots are touched up with a darker brown, to give depth to these false pores. When it is explained that *D. johnstoni* feeds on sponges like *Halichondria panicea*, this colouring is easily understood, but its marvellous nature is not lessened.

Some species of allied genera are quite remarkable, one might almost say eccentric, in their ornamentation. *Ægirus punctilucens*, a species found between tide-marks, is elaborately covered with large tubercles and shining points; the branchial tufts assuming quite a tree-like growth in miniature, around the orifice, which is placed further forward than in *Doris*.

The Crowned Eolis (*Eolis coronata*) has a slender body, long slender tentacles, that cannot be withdrawn, and the back is covered with long papillæ, gathered into a dozen

spreading bunches. The two *erect* tentacles behind the long
pointed pair, if examined with a lens, will be found to be
beautifully ornamented by a series of annular plates. It may
be sought among the rocks at low-water, feeding chiefly on
the sertularian zoophytes. It is an active species, gliding over
the rocks, or swimming at the surface with its back downwards.
They are constantly waving their tentacles and moving their
papillæ, from which they exude a milky fluid when irritated,
and even throw them off, as a crab "shoots" his lesser limbs

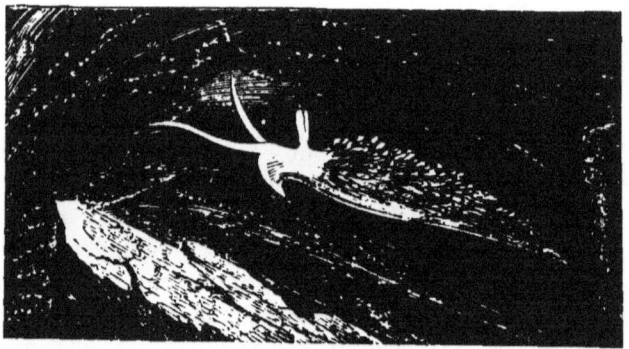

CROWNED EOLIS.

under similar circumstances. If kept in an aquarium without
suitable food, they become cannibals. *Eolis papillosa* is a
similar species, the upper surface almost completely covered
with papillæ. It will be found under stones at low-water, feeding on *Botrylli* and other ascidians. If on a white species,
it will be wholly white, for like *Lamellaria* and *Doris*, it goes
in for protective colouring. Introduce a specimen from a white
ascidian into a vessel containing, say, a crimson or brown
Beadlet Anemone, and after a few hours you will find the
Anemone has disappeared, whilst the Eolis has changed to
the colour that the Beadlet was of. The papillæ of the Eolis
are really continuations of its digestive apparatus, and by this
simple arrangement a protective harmony is set up as often

as it may change its diet. Scientifically these papillæ are termed *cerata*.

The last of our Sea-slugs does not belong to the Nudibranchiata, for its branchiæ are concealed, and it possesses a shell —a thin, flexible, translucent, convex plate, that covers the

SEA-HARE.

branchial plume, and is itself covered by the mantle. My first Sea Hare (*Aplysia depilans*), was taken in ignorance. A hurried glance at a globular mass of purple-brown jelly, among some small weeds, as I was hunting for anemones, assured me I had something new to me, and I put it down at the moment as a colony of compound ascidians; but on putting it into an aquarium, I saw my mistake at once. The bundle unrolled, and some loose wraps, shaking themselves out, resolved themselves into tentacles and marginal lobes. The foot lengthened out, and I saw the creature had a distinct neck, with a broad muzzle between the first pair of tentacles. The second pair were folded, so as to present a strong suggestion of the ears of a hare, and this is precisely the idea suggested to fishermen in many countries, by whom the *Aplysia* was first called Sea-hare, or *Lepus marinus.* When it is viewed from the front, as in the smaller illustration, the illusion is strengthened. It has the habit of pouring out a violet fluid from the edge of the mantle when handled, which is probably intended, like the Sepia's ink, to produce a cloud, under cover of which the Sea-hare can safely retreat. In other days, this fluid was regarded with horror as a poison, and an indelible stain. From this last notion the creature got its name, *Aplysia*, which is from two Greek words, meaning unwashable, filthy. Its second name,

SEA-HARE, FRONT VIEW.

depilans, is also reminiscent of those old notions, for it was thought that mere contact with the dreaded creature would cause the hair to fall off. The Sea-hare of the present generation, however, is quite harmless, as I can testify, whatever may have been the real or assumed character of his ancestors.

CHAPTER XV.

CUTTLES.

THE old trouble about a name crops up again. We have had to endure star-fish, jelly-fish, shell-fish, and now there remains cuttle—no, we will not say cuttle-*fish*. My neigh-bours, the brave Cornish fishermen, do not say the word, neither will I. With them it is " cuddle," with me it shall be Cuttle, Squid, Octopus, and so forth.

The term Gasteropoda has been explained as comprising those mollusks whose belly is also their locomotive base, so it will be easy to show that the class Cephalopoda consists of those mollusks whose feet (*tentacles*) are ranged round their head (Greek, *Kephale*, head, and *poda*, feet). They are the most highly developed of all the mollusca, and consequently come nearest to the back-boned animals (*vertebrata*). In them we find the first form of a skull, for the nervous system is more concentrated, and the brain has a cartilaginous cover-ing. The head is distinct, and there are two large and prominent though stalkless eyes. The jaws are powerful, and these work in a similar manner to the bill of a bird. There is a thick, fleshy tongue partly covered with hooks for tearing flesh. The round or elongated body has usually a flap on each side, which serve the office of fins. The respiratory apparatus consists of two plume-like gills, enclosed in a large branchial cavity, communicating with the outer waters by a siphon or funnel. They also possess a bag of reliable black ink, of so readily soluble and miscible a character that a little ejected through a special duct will raise a dense cloud in the water with great rapidity, and under its cover the cuttle can quickly disappear. The tentacles number eight in some species, ten in others, and they are studded with a great number of suckers, which appear to be set to work almost automatically on coming into contact with any animal substance, to which they adhere so perfectly that, unless the will of the animal interposes to release their hold, it is easier to tear off the tentacle from the cuttle's body than to separate it from its victim.

The Cuttles cannot strictly be called shore creatures, but they are very active, and come into every zone, the littoral as

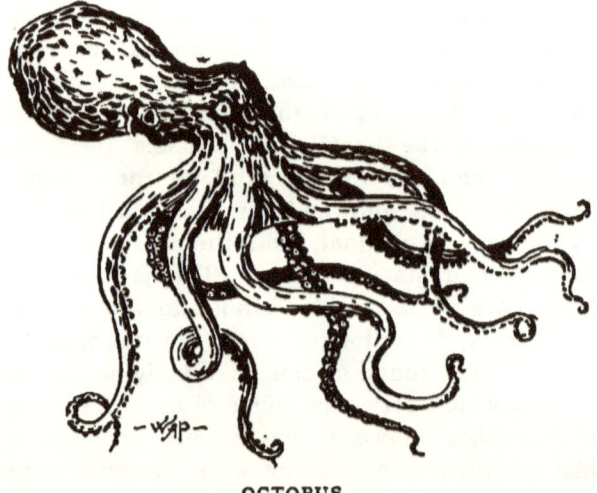

OCTOPUS.

well as others; and though we are not very likely to come across the animal itself, we are sure to find Cuttle-"bones" upon the beach, and bunches of their eggs. In our investigations of the rocks at low water, we may perchance come across a specimen of the Octopus, hiding in its hole under the weeds, or beneath a big stone we have just overturned. Occasionally, too, it may be found in a pool that is covered by a fathom or so of water at ordinary low tides. On being discovered, it immediately, and with considerable force, ejects a stream of water through its syphon from the branchial chamber, and by the recoil is sent backward through the water. As it does so we can see the play of colour over its body, showing that the pigment-cells are ever ready to accommodate themselves to the surroundings, that the Cuttle's skin may imitate them. It is not very likely to discharge its inky cloud, for *Octopus vulgaris* is not so ready as other species to empty its ink-bag, and the ink is of a thicker, less soluble nature.

The principal food of the Octopus appears to be the smaller crustacea, and their hunting period after sunset. This is the reason why so common an animal is so little seen. The shell is represented in the Octopus by two short rods of shelly matter embedded in the mantle. The body is like a round-bottomed bag, there being no side expansions (so-called *fins*), and the arms are connected by a web at their base, the suckers in two rows. The eyes fixed and staring.

Much more in evidence as a shore animal is the Sepia, the true Cuttle (*Sepia officinalis*), which lives in shallow water, and whose egg-clusters and broad internal shell we frequently encounter on the beach. The Octopus has but eight arms all told; the Sepia is adorned with other two, but these are differ-ent from the eight, and may be more correctly distinguished as tentacles. They are much longer than the Sepia's body, very narrow, and without suckers, except near their free ends, where they expand considerably. The outline of the body, apart from the head and arms, is like that of a shield with pointed base. There are narrow expansions right along the sides, serving as fins, the suckers are stalked, and the large eyes are moveable in their sockets. There are four rows of suckers on each arm, and the arms are short. The shell is the familiar "Cuttle-bone" sold by bird dealers, to provide im-prisoned songsters with the necessary lime, and by chemists to be pounded and used as a dentifrice. These shells are familiar to all, and need not be described. They are exceedingly light for their size, one of average proportions (7¼ by 2½ inches) weighing less than one ounce. This is the average of the large shells one finds upon the beach, but a full-sized one would be about ten inches in length. It is technically known as the *sepiostaire*, but "Cuttle-shell" (not "bone") is good enough for common use. It should be observed that this shell serves as a complete shield for the back of the Sepia, it being merely covered by the mantle, to which, however, it is not attached. Besides its value as a shield to the Sepia, it is also useful as a float, for the Sepia is an active swimming creature, not a crawler on the sea-bottom like the Octopus.

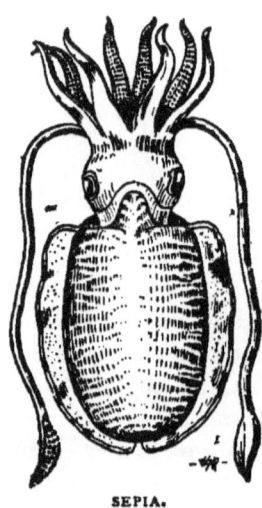

SEPIA.

The Sepia's ink-bag must not be forgotten; you are not likely to forget it if you capture a Cuttle. On one occasion when I had been out in the sean-boat capturing mackerel, I saw several Sepias swimming about among the imprisoned fish, and a couple of these contrived to be dipped up in the tucking mound, and cast into the boat with the fish. One of these I claimed as part of my share, but when we landed the creature was in such a mess with his own spilt ink that I essayed to wash him in a pool. I soon tired of that, for the more I washed, the more freely the ink was poured out. The Sepia sometimes visits the fish-nets and seans in shoals, and does great damage to the catch; but fish are equally fond of Sepia, and if you can get hold of a couple of these, or of Squid, on starting for a fishing excursion, to cut up for bait, you will scarcely want anything better. The Sepia's eggs, in clusters not unlike bunches of grapes, are frequently cast up on the shore by storms, and there is no great difficulty in hatching out such of the eggs as have not been injured by the buffeting against rocks and shingle they have experienced. The young Cuttle is a miniature *replica* of its parent, and conducts itself as "a chip of the old block."

The Squid (*Loligo vulgaris*) is a much longer and narrower species of Cuttle, similar to the Sepia in its head parts, but the arms have but two rows of suckers on each, though the clubbed ends of the tentacles have four rows. The fins are short and angular, placed at the hinder end of the body, which runs off to a long sharp point behind them. The shell is not a broad expansion like that of Sepia, but more like a pen with a long holder or shaft in front of it. Whilst the Squids are splendid swimmers, they also crawl, head downwards. This is the species that is chiefly sought for bait, and vast numbers are used in the Newfoundland Cod-fishery.

SQUID (*loligo*).

There are a number of species of Cephalopods to be caught off our coasts, but the only other that we are likely to find any trace of upon the shore is the Little Cuttle (*Sepiola rondeletii*), whose body is short, with rounded side fins, contracted at their base, and whose entire length is only a couple of inches. The suckers are in two rows on the arms, and in four rows on the tentacles; in this respect it agrees with *Loligo*, to which it is much more nearly related than to *Sepia*. It is a very active swimmer, and it has a small pen similar to that of *Loligo*.

CHAPTER XVI.

SEA SQUIRTS.

THE other day I was down in our porth when some of the fishermen of the village came in after hauling their trammels. There had been a "good bit of sea" running during the night, and the trammel had got fairly filled with weed, so that it was necessary to bring it ashore to clean it. If the naturalist is about when this happens, he stands a chance of obtaining some deep-water specimens of interest to him. My eye fell upon several masses of a clear greenish-white jelly, pear-shaped, and firm to the touch. I knew what they were, but always anxious to get local names for natural objects where they exist, I asked the fisherman what they were. "Oh, I dare say you know, sir; but we always call *they* congealed water. Isn't that right?" I admitted that they were composed almost entirely of water, but denied that it was congealed. It would be better, I added, to speak of it as a living leather bottle full of water—and other things.

"What was it?"

Popularly speaking, it was a Sea Squirt. A naturalist would speak of it as a simple *Ascidian—A. mentula*, to wit; and on being further pressed, might tell you that the Ascidiaceæ constitute an order of the Tunicata, which is now included among Vertebrate animals, though no Tunicate possesses a backbone.

Our description of it as a leather bottle is more to the point, and equally scientific, for the naturalist who bestowed the name Ascidian upon this remarkable group of animals got that name from the Greek word *askos*, a leathern bottle.

Look at these diagrams: they represent two common forms of Ascidians, and it will be noted that they have a general agreement in shape with the large specimens of *A. mentula* we were looking at just now. Like that, these have each two necks, though those of *mentula* were closed, and these are open at their mouths. If we had these in a glass vessel, but still attached to pieces of the rock upon which they grew, we should be able to see why one bottle need have two necks. If

A. ASCIDIA VIRGINEA.
B. CYNTHIA QUADRANGULARIS.

we were then to drop a little finely-divided colour-powder such as indigo, into the water, we should see two currents were in operation, one flowing to the animal, the other proceeding from it. The first would be flowing to the neck marked *a* in the figures, and the second would be issuing from the mouth of *b*. Naturally, we should at once suppose that by means of some internal mechanism and system of valves, the same current that was being induced at *a*, was being continued through the creature's body, and pumped out at *b*. Our supposition would be proved correct by the fact that the colour grains streaming in were also streaming out. But what happens to them between entering and departing we cannot clearly see.

By the aid of another diagram (next page) we may get a better notion of the Ascidian's internal arrangements than by gazing through its integuments. Here are all its parts marked with a letter as a guide to its anatomy. It is a matter of astonishment to many fairly intelligent people, to find that such soft creatures as Sea-squirts, Jelly-fishes, Slugs and Caterpillars, are provided with a more or less intricate machinery for carrying out all the functions of life. But so it is; and here is the typical plan of arrangements inside our Ascidian. Here the necks of the bottle are marked *a* and *n* respectively, and *a*, by which the current of water flows in, is called the oral orifice. Just inside is a series of tentacles (*b*), and below these we are in the branchial chamber (*c*), where the great work of supplying the blood with oxygen is carried on. The walls consist of a lattice-work of blood-vessels, through whose tissues the blood takes up the molecules of the life-supporting gas. Below this chamber the gullet opens and is continued into the stomach (*g*), and beyond it is the intestine (*h*), which in turn opens out through the anus (*l*) into another roomy chamber, the *atrium* (*m*) or atrial chamber, with its external opening (*n*). *O* is a ganglion or small brain, and *f* indicates the heart.

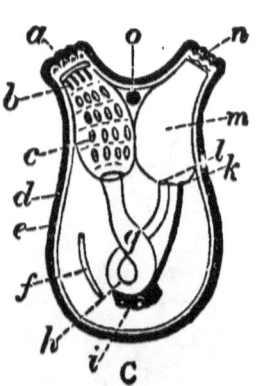

DIAGRAMMATIC SECTION OF. AN ASCIDIAN.

Now in order to get a correct idea of the Tunicates—as the group in which the Ascidians are included is called—I wish you to note the figures *d* and *e* in the same diagram. You will see that they indicate two separate envelopes. The outer of these, represented by the thick dark line, is of a tough, leathery nature, and is much akin to vegetable cellulose in its character—a fact that caused some little commotion in scientific circles years ago, when it was first satisfactorily made out, for prior to that date cellulose was considered to be purely a vegetable product. This outer coat is known as the *tunic*, or test, and from the fact that all the species are enclosed in such a tunic, the group gets its name Tunicata. The inner coat represented in the diagram by the clear space between the thick and thin marginal lines, is of a more delicate, more animal nature: it is composed of soft though powerful muscular tissue, and by its contraction the water, which always fills the interior of this " leather bottle," can be violently spurted forth—a phenomenon which has brought upon these creatures the name of Sea-squirts. This muscular coat is known as the *mantle*.

The Ascidian has no proper system of blood-vessels, as we are generally acquainted with them in higher animals. The

ORANGE-SPOTTED SQUIRT (*Cynthia aggregata*).

blood flows about the general body cavity, and is not confined to narrow channels as in our arteries and veins. There is a heart, it is true, but one of the simplest character, without any

Q

elaborate system of ventricles and auricles, with their regula-
ting valves. The Ascidian's heart is simply a tube open at
each end, and by its steady pulsation—that is, its alternate
contraction and expansion, it sets the blood flowing to the
blood-vessels that line the walls of the branchial cavity, where
it absorbs oxygen from the continuous flow of fresh sea-water
that passes through it. When this end has been attained, a
curious and unique "reversal of the engine" takes place:

Ascidia mentula.

CURRANT-SQUIRTER (STYELA GROSSULARIA).

there is such an opposite action of the heart, that all this
vivified blood is withdrawn from the neighbourhood of the
branchial chamber and sent flowing to remote parts of the
body.

The flow of water through the branchial chamber is kept up by the regular and unceasing lashing of eye-lash-like *cilia*, with which the blood-vessels are fringed. This constant inflow at the oral orifice forces the water through to the atrial chamber, from which it is pumped out by the contraction of the mantle. Minute particles of matter that serve as food are also brought in by the current, and find their way into special grooves for their reception and digestion. The other arrangements of the creature are equally simple. The nervous ganglion, to which we have made reference, is its only brain, and it has no proper eyes, only some pigment granules near the tentacles appear to be sensitive to light.

Most of the Ascidians inhabit deeper water than comes within our range, but we shall find specimens at low-water attached to stones and the roots of seaweeds. We may even find specimens of *Ascidia mentula* in rock-pools, and others we shall discover on smaller stones and shells that have washed in on sandy shores from greater depths. Among such will be the Quadrangular-squirter (*Cynthia quadrangularis*), so-called on account of the squareness of its apertures; and the Currant-squirter (*Styela grossularia*), a very common form on dead shells, which gets its name partly from its colour and partly from its form when it has closed both apertures and become more rounded.

But there are many other forms of Tunicates that haunt our shores either in deep water or upon the vegetation of the lower rocks. There are some of more slender, more elongated form that live together in bunches, their bases being connected by a kind of running rootstock, which has the power to produce young individuals by budding from it. This form is known as *Clavelina lepadiformis*, and is only about an inch in height, of the form shown in the

CLAVELINA.

annexed diagram. In the figure the reference letters are of
the following signification : *a*, branchial apertures ; *b*, atrial
apertures ; *c*, young individuals arising from the runners *s*.

From this form it is an easy transition to the Ascidians
known as *Salpæ*. These have the branchial aperture (*b*) at

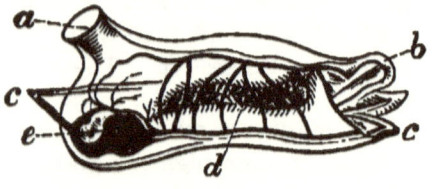

one end, and the atrial
opening (*a*) at the other.
In the figure the heart
is shown at *e*, and the
branchial chamber at *d*.
These *Salpæ* are both
solitary and compound
Ascidians, for it is a

SALPA MAXIMA.

singular fact that the solitary form as here shown produces
buds which develop into a connected series or chain of individuals. These, in turn, instead of reproducing the species,
in a similar manner produce eggs, each of which gives rise to
a solitary individual. In our figure of *Salpa maxima*, the
letters *c* indicate the points of attachment of the Salpa colony ;
and the next figure represents a portion of the Salpa-chain.

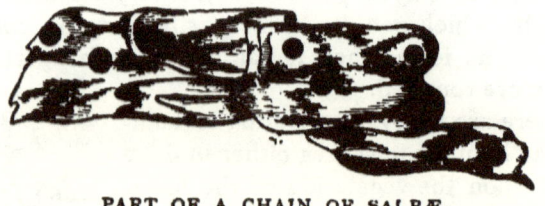

PART OF A CHAIN OF SALPÆ.

Frequently, in gazing down the sides of a still, deep rock-
pool, we shall observe a coating of dark-grey jelly, in patches
as big as one's hand, and on carefully taking off an inch or
two of this, and examining it with the lens, or a low power of
the microscope, we shall observe a number of elegantly-formed
jars to be set in the jelly, and as we look their mouths and
necks are raised above the surface of the jelly and opened.

These are the branchial apertures of a colony of Ascidians (*Leptoclinum gelatinosum*), and if we search around the mass we shall shortly find a cone-shaped opening in the clear jelly, through which a current of water flows. This is the common atrial chamber of the whole colony. The clear jelly is the common outer tunic of the whole community.

On the walls of overhanging rocks, at low-water, many fleshy clusters, like pale-coloured strawberries will be found, of firm gelatinous material, with a clear jelly envelope, through which the crimson dots of the contained squirts may be seen. One form has a thick trunk, with but slightly enlarged head, and consisting of a number of groups of squirts : this is *Aplidium;* it has no common aperture. A more globose head on a shorter stalk has a distinctly marked common opening : this is *Polyclinum*. *Amaroecium* has a corrugated exterior, and is more cylindrical in form.

BOTRYLLUS.

Other species will meet us of more symmetrical form, on flat weeds, smooth stones, and under the overhanging brows of the large rocks at low-water. These are of varied tints according to species, but each with a starry pattern worked in with little purple or yellow Ascidians. It looks as though six or seven of these had agreed to live together for company's sake and for economy ; and here we find them set in the jelly, and radiating from a central aperture, the common atrial opening of the colony.

Here is a figure showing part of a patch of *Botryllus violaceus*, such as you may find abundant on the rocks. *C* shows the combined tunic of the colony, *a* the branchial openings, and *b* the common atrium.

BOTRYLLUS VIOLACEUS.

The general verdict on a patch of Botryllus would probably be that it was some low form of sea-plant, for a naked-eye view of it reveals no evidence of animal processes ; yet, in spite of its vegetative condition, this—in common with other Tunicates—is held to approach nearest to the great back-boned races, the aristocracy of animal life.

But it is a sad story of missed opportunities and consequent degeneration that the Tunicates have to tell of their race. Some evolutionists hold that in the primeval Ascidian we must look for the progenitor of the vertebrates. We know what the primeval Ascidian was like, for the form is retained, according to a natural law, in the larval stage of its present-day representatives. Roughly speaking, it was like a tadpole, with a broad head-and-trunk combined, and a very long, narrow tail, by the lashing of which from side to side it made way through the waters, much as the boatman gets along by sculling from the stern. At the front there was a rudimentary mouth with three suckers, an optic organ, with a retina, lens, cornea, and so forth ; an auditory organ ; the promise of a well-formed brain and nervous system ; and a rod in the tail might be developed into that backbone which is the distinguishing mark of all the birds, beasts, fishes, reptiles, and man himself.

LARVA OF A TUNICATE.

Some of the primeval squirt-larvæ are supposed to have cultivated these possibilities, and the grand vertebrate division of the animal kingdom is said to be the result; but others went in for the *status quo* and inglorious ease. No developments for us, said they. They may even be supposed to have anticipated the prayer formerly taught to rural school children:

"God bless the squire and his relations,
    And keep us in our proper stations."

Then they gave up wandering at random through the waters, and settled down to a quiet and retired life on a piece of rock at the root of a branching weed. Taking hold with their suckers, they soon discovered that tails and sense organs were of no use to those who had forsworn wandering, so they threw them off, and gradually assumed the wine-skin shape that has ever since been the ruling fashion among Ascidians. All that remains of the tail is a few fatty cells in the posterior part of the trunk. The suckers by which it was attached disappear, and the test grows over surrounding objects; the auditory organ disappears, the eye retrogrades into a mere pigment spot, and the nervous system degenerates into the solitary ganglion to which we have already referred. It will thus be seen that the life history of the Tunicates is a dismal story of degeneration instead of development; but it is none the less interesting on that account.

CHAPTER XVII.

SHORE FISHES.

WE have no intention of attempting to give in this little book an account of British marine fishes. That is a task that needs several volumes for its accomplishment. But without going from the shore we may make acquaintance with a considerable number of fishes. Where trawlers come in we may, of course, see fish of all sorts, but as in most cases the trawlers put in with their catch to the nearest market-port, we shall take no account of this method of increasing our knowledge. From time to time the local fishermen get strange things in their trammels, such as enormous angler-fishes; one day one of our fishermen got a porpoise in this way, and brought it ashore for my special benefit. But these things also I shall treat as outside our bounds, which includes only the fish we can find in the rock-pools, or under stones at low-water, or can catch from the fringing rocks as they haunt the weedy jungles of such places.

To begin, let us take some fair-sized rock-pool, between tide-marks; one with irregular walls overgrown with green and purple weeds, and pinkish coralline—with miniature caverns and clefts in the walls, and a heavy stone or two at the bottom. In such a pool—and we know hundreds such—we shall not fail for several examples of fish, though we are not likely to find all the species here named in one and the same pool. Three or four species of fish at the most is what we may expect from one pool; but in several basins within a few yards of each other we may get a greater variety.

In all probability the first species we shall see in the pool is the Smooth Blenny or Shanny (*Blennius pholis*), which the boys in my neighbourhood (South Cornwall) call Janny, and in other districts it is the Mulligranoc. It is a true rock-fish, never venturing into very deep water, and preferring those pools between tide-marks where it can find convenient shelter in holes, or if so inclined can climb out and pass a few hours under the moist weeds which the ebbing tide has left uncovered. But it is never many inches from the water, and on the least sign of alarm it is in the pool and invisible.

In many respects it is a clumsy, heavy fish, but its quick intelligence makes up for defects of form—and we may add, makes it an interesting fish to keep in a shallow pan with a few stones. You must have the stones if you would have the

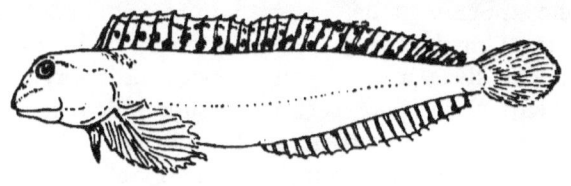

SHANNY.

Shanny comfortable, for he is strongly averse to too much publicity. He likes to see and not be seen; and his favourite attitude, so far as I have observed a number of specimens in confinement, is on his side under a stone, with the head just peeping out. In this position he appears to have one eye on the floor of the tank, the other on the surface of the water. Look at him and he follows your every movement with one eye. In this position he reminds me strongly of a dog; indeed, in certain aspects of his profile his head much resembles that of a dog. He acts like a dog, too, when he has taken a limpet unawares, and has wrested it from the rock. This is not an easy thing for a fish to do, and you might almost as well speak of taking a limpet off-guard as of catching a weasel asleep. But for some reason—perhaps to thoroughly ventilate his shell, or for the submarine equivalent for ventilation—the limpet occasionally lifts his shell so that there is about an eighth of an inch clear space between the edges of his shell and the rock. He still retains his hold by means of his powerful sucker-foot, but the wily Shanny, creeping silently up seizes the shell in his strong lips, and before the limpet can exercise his muscular powers by pulling down the shell and pinching the shanny's lower jaw, the fish, with a shake of his head, has wrested it off the rock. He carries it about for some time, biting at the flesh and gradually reducing it in quantity.

Each Shanny occupies his own private corner or crevice of the pool and shuns the company of his fellows. In this matter he appears to be a very morose fish, and further he resents anything in the way of a friendly call. Should the Shanny, who lives in the grotto about half way along the southern side of the pool, seek to call upon his neighbour who lives in that delightful retreat at the bottom of the west end, the latter will rush out at him like a mad bull and effectually put the visitor to the rout.

In every pool there are a number of juvenile Shannies of various ages and sizes, but of these the adults do not appear to take much notice. One of the most noteworthy things about the Shanny—shared I admit with many other fish, but still worthy of observation—is the rapidity with which he can make himself practically invisible. It is not easy to describe the Shanny's coloration and markings, because it varies so much in different individuals, and even in the same individual at different times; but it may be said to be a mottling of greenish-grey or brown marks, of which the strongest elements are a series of dark broad stripes, running from the back to about half way down the sides. The whole of the upper surfaces are liberally sprinkled with small black or grey dots, and larger ones are scattered over the dorsal fin, which is continuous from above the gills right along almost to the tail, which is similarly spotted. There are very few spots on the long anal fin which is hidden when the fish is resting; but the expansive oval pectoral fins, which are often spread out widely, have the rays well-spotted.

You lift up a big stone from the bottom of a pool and out rushes a big Shanny, causing a great commotion in the water. He makes for a narrow cleft where there does not seem to be nearly sufficient room for so big a fish as he; but he has vanished. Knowing where he disappeared you rout him out again, and once more he frantically flies round and round the pool, perhaps leaping right out of the water into a tuft of overhanging *Fucus serratus.* But as likely as not, after dodging about for two or three turns and splashing the water about, he will quietly drop to the coralline-covered floor right under your eyes, and you cannot see him. So admirably does the indefinite marking of his upper surface harmonise with the coralline and other matters, that he has become as invisible as a nightjar on the moorland, or as certain moths on lichen-covered tree-trunks. It will do you no harm to carefully scrutinise every millimetre of the pool's floor until you have detected the Shanny's whereabouts, but probably you will be assisted in this by the Shanny himself, who, observing your quietness, will imagine all danger is past and make a move.

Juvenile Shannies, though as ready to rush into cover as their elders, are endowed with considerable curiosity; and if in early summer you come upon a dozen of them sporting about a rock-pool, and will lie down with your head and shoulders over the water, you will find that their inquisitiveness is greater than their fear. One after another will come from his retreat among the weeds and look up at you, rolling his little eyes knowingly. Then they will creep up the sides of the pool, using their ventral fins as feet, until their muzzles are out of water. Dip in the tip of a finger, and they all vanish for a moment; then out they come again, and slowly approach until they reach your finger; they attempt to bite it, but their mouths are as yet too small, and then rush off again. So you may keep them employed for some time, and it will not be many minutes before several prawns join in the fun. This may read like an ordinary "fish-story," but it is a fact that may be verified by any visitor to a rocky shore.

Next to the Shanny we shall probably find the most reliable fish as a pool *habitué* is the Father Lasher, Horny Cobbler, or Sting-fish (*Cottus scorpio*). Put but the point of a stick in the pool where the Father Lasher has his retreat under a stone, or drop a winkle or a pebble in; in an instant he is out with open mouth ready to swallow anything not too large for his very capacious maw. His singular name appears to have been given to him on account of his pugnacity and the villainous expression of his countenance, which are supposed to belong to a creature who would not hesitate to give his own parent a thrashing. My own opinion, based upon considerable personal acquaintance with the Father Lasher is, that he is not nearly so villainous as he looks. His case is similar to that of the bull-dog, whose face is no index to the qualities of heart I am told he possesses. The artists have not been fortunate in depicting the Father Lasher, and I am not greatly surprised, for even the camera fails to give a correct and life-like impression of him, which depends not alone upon curves and lines, but upon colour also.

In some respects he resembles the Shanny in build, but is much broader across the head and shoulders. He has the same wealth of fins, though the dorsal fins are not continuous as in the Shanny, and the fin rays though stout are soft. There is an inclination towards the tadpole form, especially on the under side, and this tendency is exaggerated by the fish puffing out his gills and sticking out his pectoral fins when threatened or alarmed—or when he wishes to inspire with awe. Just behind each eye and at the top of each gill-cover he has

WORM PIPE-FISH.

FATHER LASHER.

a bony spine, with smaller ones all over his head, and the inflation of his jaws and gills is for the purpose of forcing these out. Whether he makes any use of them in actual warfare I am unable to state, but they certainly add to his ferocious aspect, and in that way may protect him from many assaults. The more barbarous of the coast-boys delight in fixing corks to these spines, and setting Father Lasher free, get amusement out of his vain efforts to seek his hole at the bottom of the pool.

The Father Lasher's colouring is a confusion of bands and circles and spots; of browns and greens and greys; a serviceable coat that harmonises well with all its surroundings, and one that is capable of adaptation when the fish moves from a bare rock basin to one that has a coralline lining. It can change from dark to light, or *vice versâ*. I have had them almost white by keeping them in a white porcelain dish. The underside is delicate yellow, or pearly white, or iridescent green with darker mottlings.

In his native pool the Father Lasher likes to take up his quarters under a stone at the bottom, from which he can suddenly rush out at anything he sees move across his field of vision. He does not wait to see what it is; sufficient that it moves. Satisfied that movement is a sign of life, he secures it in his cavern-like mouth, and then finds out whether it is a palatable morsel or not; if not, it is summarily ejected and, as he thinks, no harm is done. An angler who simply desires sport can get it in a pool where lives a Father Lasher. Drop down a baited hook, and it will soon be seized by him, but, as he immediately retires to his den to chew it over, you may pull and pull before you get him out. Probably you will lose several hooks before you secure your fish. He is not at all a bad subject for an aquarium proportioned to his size, and he soon becomes quite affable, allowing himself to be taken out to exhibit the beautiful marbling of his underside to friends. For this purpose I have held him gently with my finger and thumb behind his pectoral fins, when he would obligingly open his enormous mouth to show how well the jaws and palate are furnished with teeth. When fully grown he attains a length of five or six inches.

Our illustration on page 251 contains a portrait of the long and slender Worm Pipe-fish (*Syngnathus lumbriciformis*), besides that of the Father Lasher. A more striking contrast could not be desired between fishes of the same length, for the Father Lasher is thick and spiny, whilst the Worm Pipe-fish almost comes within the definition of a line, "length without breadth," and in addition he is as smooth as an eel, though of harder exterior. This little fellow might more easily pass muster as a worm than as a fish. It will more frequently be found under stones at low-water, but occasionally we shall find it in the pool twining S-shaped round some seaweed.

The peculiarity of the pipe-fishes, of which we have several native species, is to have these long tapering bodies, with the snout drawn out into the form of a beak, but which instead of separating into two mandibles, opens only at the extremity with a little mouth. Another distinguishing feature is found in the gills: instead of these being a series of crimson frills covered by a large plate, fixed only by a small portion of its edge, and freely opening to allow the passage of water to and from them, their blood-vessels are gathered into little tufts which are arranged in pairs. These are all covered in by a bony plate that is fixed all round, with the exception of a small opening near the top edge. Then instead of the body being covered with scales as in many, or most, fishes, these are encased in large plates of mail. In the male of our Greater Pipe-fish or Greater Sea Adder (*S. acus*), there is another remarkable item in the shape of a marsupial pouch of the same practical value as that of the Kangaroo, into which the female transfers her eggs, and where they not only remain until they are hatched, but the young fish also use it as a shelter for a time, coming home unfailingly to roost. This is a fish that may be taken freely among the weeds of bays and harbours, and as it reaches a length of from twelve to fifteen inches, it is a giant compared with the little Worm Pipe-fish.

The Worm Pipe-fish has no fins except that along the back (*dorsal*), and its tail-fin is almost non-existent; it can, however, be found by looking for it. It has no marsupial pouch, but the female contrives to transfer her eggs to the abdomen of the male, where each sinks into a little pit in which it is held until hatched. How this is accomplished I have not observed; but as I have found the strings of ova independently in my aquaria, I should suppose the male presses his body upon them until they adhere. These eggs are one millimetre in diameter, amber-coloured, and opalescent. They are firmly attached together in rows of twos or threes, and these rows in circular strings. They are firm to the touch and not at all adhesive, so the glutinous matter, necessary for their adhesion to the male, must be contributed by that parent. It is interesting to note that when these tiny creatures leave the egg the tail has a proper broad fin at its extremity and extending along both the back and underside. It has also pectoral fins; but all these except a part of that along the back become absorbed, or are otherwise got rid of as the fish grows and becomes more worm-like. So smooth and round is this species that it presents little evidence of being clothed in plates instead of scales, until one looks very closely, when the outlines of each plate will be found indicated.

If the Worm Pipe-fish be captured with care, and soon transferred to the aquarium, it will be found quite a hardy and interesting inhabitant. Of course, its comfort must be studied, and to this end you must provide a flat stone, so propped up that it is very close to the bottom of the tank, yet with sufficient space beneath for the Pipe-fish to wriggle about. I write these notes with such an arrangement before

R

me, and as I look down through the shallow water I see five slender cylinders protruding like the barrels of tiny rifles from an ambuscade. Couch makes the extraordinary statement that, "observation seems to show that it is not able to raise itself above the ground, on which it creeps in its endeavours to escape being caught, with a serpentine motion much like that of a slow-worm." Observation in my case serves to controvert Couch. It certainly prefers to remain under stones, and it is not constructed as a constant swimmer; but it does swim for short lengths in its pursuit of minute crustaceans, and can be very active when it pleases.

There are other blennies in the pool besides that one called the Smooth Blenny or Shanny, and among those that we fancy are young Shannies we may chance to find Montagu's Blenny (*Blennius galerita*), a species easily distinguished by a crimson crest with fringed edges, which it erects on its head just above the eyes. Its tail and its pectoral fins, too, are tinged with crimson. Another Blenny, though by no means so likely to be found generally distributed along the coast is the striking Butterfly Blenny (*Blennius ocellaris*). It is much like the Shanny, but with larger and more rounded pectoral fins, and a much higher dorsal fin. This fin is the feature that at once enables us to identify the Butterfly among Blennies. It is often divided by one or two depressions, so that it appears to be two or three fins; but the important sign is a large deep blue spot surrounded by a light ring over the centre of the body. This eye-spot gives it the specific name *ocellaris*. It should also be noted that the first ray of this dorsal fin is considerably longer than the membranous portion of the fin. The colour of the fish is olive mottled with brown, but of course it varies considerably like the species we have already described. Ocellaris has two little crests upon its head similar to Montagu's Blenny, and the Tompot as afterwards mentioned.

In the illustration of the Butterfly Blenny there is a portrait

of a little rock-fish, one of the numerous tribe of Gobies.
Several of them occur in the pools, among them the Rock
Goby (*Gobius niger*), or Black Goby, as he is more often but
inappropiately named, for he can scarcely be said to have any
permanent colour when his hues constantly change as he
changes his surroundings. Living among rocks he is more

LITTLE GOBY.   BUTTERFLY BLENNY.

often brown than black, with lighter and darker mottlings
according to circumstances.

The reader is advised to make himself acquainted with the
names of the various fins, and to count the rays in each, for
these vary with the species, and are often used in describing

and identifying species. We have introduced the names of these already, but we think it would be an advantage to repeat them here, and then to use them throughout the remainder of this chapter.

The Dorsal fin is on the back ; if more than one they are first dorsal and second dorsal.
The Pectoral fins are a pair having their origin just behind the gills.
The Ventral fins are a pair on the belly, behind and below the pectorals.
The Anal fin is single, in the middle line of the underside between the vent and the tail.
The Caudal fin is the termination of the tail, and the form of this is very important.

The Rock Goby has two dorsals, the first with six rays decreasing in length as they get further from the head; the second with fifteen rays of equal length. The pectorals are rounded behind; so are the ventrals, which are united by a membrane. The anal fin is just under the second dorsal, if we reckon from the tail forwards, but the second dorsal is longer than the anal. The space between the dorsal and anal fins is occupied by eleven or twelve lines of scales. Full-grown specimens vary from six to nine inches in length.

The species figured in our illustration (page 257) is the Little Goby (*G. minutus*), a fish from two to three inches in length, of a yellowish ground colour minutely stippled with brown, its sides alternately streaked with long and short dark stripes. Dorsal fins two, the first rounded, narrow from back to front; the second wide from back to front, and with slightly concave outline. It appears to be more at home on the sandy than the rocky shore.

In pools that are lavishly decorated with hanging weeds we may find a number of pretty fishes of a clear green or a rich brown colour. They are the young of the Corkwing Wrasse, or Rath, as Cornish fishermen term it (*Crenilabrus melops*), a species that grows only to a length of six or seven inches. The Wrasses proper (*Labrus*), of which we shall have something to say directly, are distinguished by the oblong form of body, by having the gill-covers laid over with scales, and by

CORKWING WRASSE.

the long dorsal fin spread partly over spines and partly over soft rays. The spiny portion is the three-fifths nearest the head, the remainder being supported by soft rays. Other characters are thick, fleshy lips and protruding teeth. The Corkwing is included in the genus Crenilabrus, which is separated from Labrus on account of the margin of the first plate of the gill-covers being toothed.

In general the colouring of the Corkwing Wrasse is brown above, nicely merging into green on the sides; the gill-covers ornamented with stripes of red and green. But as we have already indicated, individuals vary much in colour. From immediately behind the head there runs parallel with the outline of the back a dark line (the *lateral line*), which terminates in a well-defined round black spot close to the tail. We must not look for large specimens, nor for the larger species of Wrasse, in the pools; but if we get on the edge of the rocks when the tide is coming in we are almost sure to see some of considerable size gliding in and out the waving fronds of the rock weeds. They are easily taken on a line cast from the rocks at this time, the hook being baited with pilchard or a piece of shore crab. Many are caught in this way for sport, and then handed over to the crabbers as bait for their pots. For this purpose they are much appreciated, and special pots are put down to capture "rath" for bait.

One of the commonest species is the Ballan Wrasse (*Labrus maculatus*), which is *the* Wrasse. The ground colour is usually some variation upon golden orange, and many of the scales have a large pale spot which earns for the species the name *maculatus*. The spines in the dorsal fin are twenty, and the soft rays ten or eleven. Certain forms are known as the Green Wrasse and the Comber Wrasse, under the impression they are distinct species, whereas they are really colour varieties of the Ballan Wrasse. The length varies in adults from fifteen inches to two feet, with a weight of eight or ten pounds.

The Cook or Cuckoo Wrasse (*Labrus mixtus*) is another common kind, not so large as the Ballan, but more striking in its vivid colouring. This varies from yellow to red as a ground tint, with two roughly parallel purple or bright blue thick lines running from above the eye nearly to the tail. The large eye is crimson with a purple ring round it, from which run off three short bands of blue or purple across the gill-covers. All the fins red, the fore part of the dorsal suffused with blue; a triangular patch of blue also on the upper and lower parts of the tail. The dorsal fin has eighteen spines and thirteen soft rays.

Should we desire to see the life of the rocks without troubling to obtain "specimens," it is a good plan to repair at low tide to the edge of a drang, and, selecting a station where we shall have a high rock in front of us and a channel between ourselves and that, wait until the tide turns. At first there is nothing but the rough floor of the drang, with stones and rocks of all sorts and sizes scattered untidily over it. The great broad, leathery fronds of oarweed and the smaller fronds of bladder-wrack and knotted-wrack hang over the rocks in great shaggy masses, and here and there, as though in utter collapse, are the flaccid forms of the green and drab Opelet Anemone. But as we are taking stock of the surroundings, there comes a ripple of water along the deeper ruts and pools of the drang. Silently it streams along filling the holes, and then gradually spreading right across the stony floor, and creeping up and up the rocks until there is an inch or more of it. Then what a change ensues. The free ends of the weeds float in the stream, the smaller weeds on the bottom pick themselves up, and shapeless masses become forms of elegance and beautiful colour. What a few minutes ago looked like the "abomination of desolation," is now full of life. The waters are teeming with forms that seem to rise out of the ground. Certainly they did not—many of them—come in with the tide. No, they were hidden in holes, under stones, under the limp weeds, and in crevices of the rock. Here they come. Prawns in shoals, little Wrasse and big ones, the long lithe forms of Gunnel and Rockling, the attenuated Fifteen-spined Stickleback, the Weever, and many another. Our attention is taken by a waving black form near at hand, and for a few minutes we are at a loss to make out what it can be. It appears to be a plant of strange nature, for it is evidently rooted at the bottom. And then a suspicion arises that the swaying and waving of the ribbon is not entirely caused by the influx of the tide, but we have not decided what it is, when up it comes with a green shore crab at the other end of it. It is a small Conger that has been struggling to bring into the light of day this crab, which it had tracked to his hole in the bottom. In such a position the crab had evidently something to cling to, but the Conger had fixed his teeth in the crab, and it was only a question of time when the crab should be unable longer to hold out. The Conger is rapidly off to his own special haunt, there to eat the crab in peace.

The Conger Eel (*Conger vulgaris*) is for its size among the most powerful of our fishes. The largest specimens, of course, are taken in deep water, but individuals of considerable size are taken from the rocks, where they have their retreats in little caverns beneath the broad fronds of *Laminaria*. Jonathan Couch remarked that he had a note of a Conger that had been taken weighing one hundred and four pounds, and of another measuring seven feet two inches which weighed ninety pounds. Even much smaller monsters than this have to be treated with caution when caught, the fishermen usually striking them a smart blow on the tail to disable them and so prevent much mischief. The upper jaw of the Conger projects over the lower one, which is the reverse of what obtains among the true eels (*Anguilla*); the dorsal fin, too, begins much nearer, and as in the eels combines with the ventral fin to form the tail.

When on the floor of the emptied drang turning stones and lifting weeds aside, we shall probably hear a great splashing in the shallow pool behind us, and turning quickly see the waters in commotion, but fail to detect the cause. But we know from former experience that it is either a Tompot, a Gunnel, or large Rockling. Fixing our eyes upon a large stone towards which the surface ripples are setting, we advance towards it and turn it over. "There he is! quick!" But no; he is as slippery as butter and glides rapidly through our hands, though not so quickly but that we could identify him as the Gunnel or Butterfish. We set out after him again, and rout him out of the corner into which he had retired in fancied safety. Next time he attempts to shelter under a stone where there is a cavity only large enough to accommodate his head and shoulders, but ostrich like, he thinks he is wholly concealed. Keeping our shrimp-net close up, we seize him just behind the head, but with a rapid turn his head is with-drawn from the hole and his body glides through our hand again, and he rushes headlong into the net. Safe this time, and soon he is transferred to the glass jam-jar where we can admire his lithe form.

The Gunnel (*Centronotus gunnellus*) looks as though by continually pushing his way through narrow crevices in the rocks, he had become laterally flattened. Were he a little rounded we might say his shape was eel-like, for he is very long, and his dorsal fin stretches from above

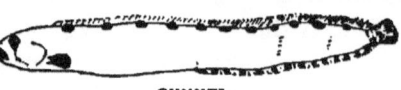

GUNNEL.

the pectoral fin along to the root of the tail. On the lower side the anal fin similarly extends to the tail, but neither of them merge into the tail-fin as in the Conger. The colour is a yellowish-brown, darker on the upper side, which is slightly mottled. Pectoral fins yellowish. Close up to the dorsal fin on each side of the back is a series of from eight to twelve— usually nine—very dark round spots, each encircled with pale brown. The head tapers gently from the dorsal fin to the small, equal jaws. It is generally known as Butterfish, and anyone who has undertaken to capture one with his hands alone will appreciate the fitness of the name, for it is so slippery that it might have been freshly greased.

Other local names for it are Swordick, in allusion to its sword-shape; and Nine-eyes, suggested by the ocelli on its back. The name by which it is best known in books is the Gunnel, which originated in a singular manner, according to Couch. It appears that John Ray, the celebrated naturalist, made his acquaintance with this fish on the Cornish coast, where it is common, and applied to a native for its name. The native was probably a fisherman, one of a class that takes little account of the inhabitants of the deep unless they are marketable sorts. He knew no more about it than John Ray did, but casting around for some analogy in the shape of the fish, he answered, "It looks like a gunwale" (pronounced "gunnel"). He thought it resembled the gunwale of a boat; but Ray naturally took "gunnel" to be the local name for the fish, and so he inscribed it in his book, and Gunnel has been the English book-name ever since, and has also been Latinized into *gunnellus* to form a scientific name.

To those who are satisfied with a cursory glance at natural objects as they flash by in life, the Rocklings might pass for Gunnels, and the Gunnel for a Rockling. The Rocklings' colour, though more ruddy and deeper, and their general form though much rounder, are sufficiently similar to warrant the superficial observer in classing them together. Their habitat, too, is much the same as the Gunnel's; and if we go down at low-water to the edge of the tide, and turn over the large flat stones that are there, we shall be sure to find a few Rock-lings of various sizes—some a foot or more in length.

Our turning-over of the stone is the signal for an excited rush, and a splashing up of the water as the Rockling dashes from stone to stone, from hole to hole. After having let him slip through our fingers, or over our hands, several times, we corner him at last, and transfer him to a large bottle in spite of his slipperiness. He proves to be the Five-bearded Rockling (*Motella mustela*), as we see at once by the four barbs on the upper jaw and the solitary one beneath. These are really sufficient for identification purposes, for no other of our shore-haunting fishes is decorated in precisely the same manner. However, we will briefly indicate the appearance of the fish. The dorsal fin commences at about one-third of the Rockling's length, reckoning from the front, and continues close up to the tail. The anal fin starts a little beyond the vent and continues near to the tail below. Pectorals rounded; ventral long and pointed. Just before the beginning of the dorsal fin there is a long, narrow, and delicate membrane that

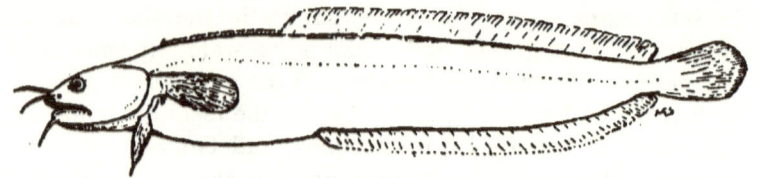

THREE-BEARDED ROCKLING.

looks like another dorsal fin, but is not. Of the barbs from which the fish gets its distinctive name, two are directed forwards and upwards, two forwards and outwards, whilst the fifth goes forwards and downwards.

The Three-bearded Rockling (*Motella vulgaris*) is very like the last-named species, but has only one pair of barbs on the upper jaw, and a single one on the lower. Mr. R. Quiller Couch discovered that one or other of the Rocklings, probably all three—for there is another species, the Four-bearded (*Motella cimbria*)—build a kind of nest by jamming fragments of coralline into a cranny, and depositing their eggs in the mass as the work proceeds.

From one nest-building fish to another is a very easy transition.

FIFTEEN-SPINED STICKLEBACK.   LESSER WEEVER

The Fifteen-spined Stickleback (*Gasterosteus spinachia*) has long been famed to be a capital builder of nests. It is at low-tide we are most likely to find this fish, though it does occasionally occur in the rock-pools higher up. We may be inclined at first sight, previous to capturing it, to regard it as one of the Pipe-fishes, already described; but its short and broad dorsal and ventral fins and its deeply-cut jaws should be sufficient to at once identify it, especially if an eye is turned to the threatening array of fifteen short stout spines that arm the back in front of the dorsal fin. Its colour varies from green to brown; in fact, the one individual changes in hue at times. It is much larger than its familiar three-spined relative of fresh-water streams and ponds, and usually attains a length of six inches. The lower jaw is longer than the upper, and both are furnished with teeth. By means of a mucus thread they are able to produce, they weave and bind together some of the softer and more delicate seaweeds, giving solidity by working in a few branching corallines, until they have elaborated a large pear-shaped mass, as big as a man's fist. In this the eggs are deposited, and thereafter are watched assiduously by the male parent, who will brook no interference, but will fiercely attack any would-be spoiler of his nursery.

Beneath the illustration of the Stickleback on page 267, there is represented the fore-portion of the Lesser Weever (*Trachinus vipera*). It is not strictly a shore fish, but it has the habit, shared by a larger relative, the Greater Weever (*T. draco*), of half-burying itself in sand and getting left dry by the ebbing tide. It *may*, therefore, be found by one of my readers in one of those spits of sand that occur between the rocks, and it is mentioned here, for the sake of warning. It must not be caught in the hands like a Shanny or Gunnel, for on the gill-covers of each side there is a long, hard, and sharp spine, which the fish knows how to use with such effect that whoever handles the Weever is likely to have a badly-injured hand. Terrible stories are told by fishermen of its effects, much of which are exaggerations; there is sufficient solid basis, however, to make Weevers undesirable acquisitions, unless we require them for museum specimens.

There is a glorious, or perhaps I should say a glorified, Blenny to be obtained sometimes at low spring-tides. It is vulgarly known as the Tompot, but it has also a literary name (by which I mean a term used only in books) borrowed from the Italian, namely, Gattorugine; and its scientific pseudonym is *Blennius gattorugine*. Its colouring is very similar to that of the Rocklings, except that it is more finely mottled and dotted. It has the general build and facial expression of the Shanny, with the crest of Montagu's Blenny. Its cheeks are full, its lips thick, eyes large and prominent. The dorsal fin begins immediately behind the head and continues right along the back to the rounded tail. The anal fin is continuous from the vent to the tail; the pectorals rounded, with fleshy rays, and the ventral reduced each to two fleshy processes with which it feels its way, as do other species of *Blennius*. It is frequently caught in crab-pots, whither it has gone for the bait, but it is in turn skewered up as bait when the crabber hauls and resets his pots. It grows to about nine inches in length. Being scaleless, like most of the rock-fishes, it is exceedingly difficult to catch with the hands.

The last of the rock-fishes to which we propose to call attention are the Suckers (*Liparis*), which must be looked for under stones at low-water. One of these is the pretty little fish represented in our illustration, and known as the Two-spotted Sucker (*Lepadogaster bimaculatus*), a species that rarely exceeds a couple of inches in length. Its head is broad and flat, the snout sharp, and the tail rounded. The lower jaw is shorter and narrower than the upper; and the dorsal and anal fins are both remarkably short, each consisting only of a few rays (five to seven) with the connecting membrane. They are placed very far back, though widely

TWO-SPOTTED SUCKER.

separated from the tail. On the underside at the broadest part of the body there is a sucking organ, consisting of a double disk united to the pectoral and ventral fins, and by means of which it quickly attaches itself to stones and other objects. The general colour is orange mottled with red; but specimens have been taken of a light brown, dotted with blue. There is a beautiful eye-like spot on each side, a little behind the pectoral fin; and it is to the pair of these that the species owes its name. It should, however, be added, that specimens taken on the shore are frequently deficient in this ornamental marking. It has a trick when at rest—and these Suckers appear to be always at rest—of throwing the hinder portion of the body round until the tail and the pectoral fin nearly touch.

A much larger species is the Cornish Sucker (*Lepadogaster gouani*), so called because it was first obtained from the Cornish coast. It is about four inches in length, of a purple or crimson tint, the under parts more inclined to pale red. Behind the eyes are two dark spots in paler rings, and with blue centres. Near the nostrils on each side are two branched thread-like processes, but of very brief length. The dorsal fin quite occupies the hinder third of the Sucker's back, and with the shorter anal fin runs right up to the tail.

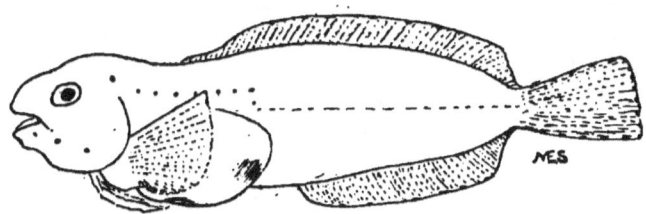

MONTAGU'S SUCKER.

Montagu's Sucker (*Liparis montagui*) belongs to another genus. It is so soft and delicate that certain yellow specimens I find attached to stones just below my study windows, look as though modelled in butter, like young Canova's butter lion. But Couch says its general colour is chestnut-brown, lighter beneath. The head is broad and flat, the cheeks chubby, the eyes small. The dorsal fin, which is marked with irregular dusky clouds, has its fore-end just above the pectorals; it is there very slight, but a quarter of an inch further back it gently rises to its full height and continues with equal depth to the tail. The anal fin is about one-third shorter than the dorsal, the curtailment being at the fore-end. The tail and some of the fin-rays are prettily dotted with black. Like the Two-spotted Sucker it reposes with its tail beside its head.

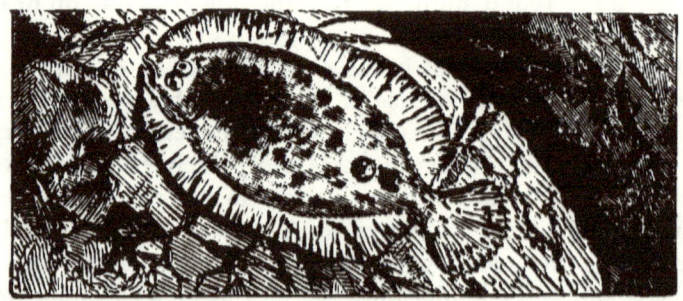

TOPKNOT.

And now for a short time we will leave these rocks, and step across to the sandy shore that spreads for a little distance round this segment of the bay. We shall not find much animal life there, except what has been washed in with the loose weeds and rubbish. You see from the absence of rocks there is no protection, and no firm basis in the ever-shifting sands for one to make a home. It is only burrowing molluscs, a crab or two, and two or three fishes that we may expect to find. Occasionally, where the water is very clear, we may see small flat-fish swimming in their strange but elegant fashion, and when wading we may chance to put a bare foot on one that is resting on the sand, where they are invisible. Among these may be the little Topknot (*Rhombus punctatus*) of our figure, and the Flounder (*Pleuronectes flesus*).

It is a common error to refer to the coloured upper surface of all flat-fishes as the back; but they are not *de*pressed, they are *com*pressed, as the position of the fins and gills should teach us. When quite young their eyes were situated one on

LESSER LAUNCE OR SAND-EEL.

each side of the dorsal line, but from their habit of resting always on one side at the bottom of the sea, the eye that is below gradually comes to the other side, so that in the adult flat-fish the pair are close together. The Skates and Rays, on the other hand, have been flattened from above; the mouth is underneath, where also are the gill openings, whilst the eyes, at a proportionate distance from each other, are placed symmetrically on the upper side.

If we take a trowel or spade we may succeed in digging up some specimens of the Lesser Launce (*Ammodytes tobianus*), often incorrectly termed the Sand-eel. He who would catch this beautiful little fish must be very quick, for if, on being dug out, it is allowed again to touch the sand, it will disappear with such speed as makes it well nigh impossible to overtake it again.   Couch says it rarely goes from the sandy shores far into deep water. It swims in small schools in the quiet waters of bays and harbours where it may be seen in summer to be chased by mackerel and other fishes.   It will be observed in the accompanying figure that the lower jaw is longer than the upper, and when the mouth is closed, the fleshy pointed edge of the under jaw furnishes a valuable instrument for piercing the soft sand when the Launce is thus beset by enemies. The outline of the fish is so evenly, gently tapering that there are no elevations that can offer resistance to its rapid progress through the sand.   It has a clean, keen look, as though the sand had been used for scouring and sharpening it.   The pectoral fins are long and narrow, but there are no ventral fins.   The dorsal fin extends almost from the pectorals to the tail at one height throughout, and the same description applies to the anal fin in its course from the vent to near the tail.

CHAPTER XVIII.

BIRDS OF THE SEA-SHORE.

ONE of the greatest charms of the sea-shore to the majority of visitors is afforded by the marine birds in their varied occupations of flying, swimming, diving, and walking. In these beautiful creatures the British coasts are rich, even when we exclude (as we propose to do from this chapter) the many species that frequent the mud-flats of estuaries in preference to the rocks and sands of the sea-coast proper. Strange as it may at first sight seem, the sea-sands are in the hard weather of winter the resort of multitudes of small birds from inland woods and commons, which here seek their sustenance at the very time when hunger induces the gulls to follow the plough and to penetrate far up the rivers—even to such uncongenial places as London itself, where, however, they are sure of a cordial welcome and a plentiful repast. Then is the season for the starlings and the thrushes to take their sea-change, and I have seen them in winter in great crowds upon the sands, hobnobbing and competing with rooks, redbreasts, lapwings, and finches of many kinds, for the odds and ends brought in by every wave, and for the smaller mollusks, the marine worms and minor crustaceans that the shore affords to the quick-eyed and the patient seeker.

But our business just now is more with those birds to whom the shore and the adjacent waters are their every-day hunting-grounds, the place where many of them lay their eggs and rear their young. One of the most constant of these is the Rock Pipit (*Anthus obscurus*), whose happy chirrup and light-hearted springy flight from sand to rock, or rock to rock, are every-day and all the year round features of the shore on certain parts of our coast. In other places it is only seen at the period of migration or in the winter months. It is a larger and darker bird than the well-known Meadow Pipit (*A. pratensis*), and its hind claw is more curved and not so long. The bill is black, with a little yellow at its base, and the tail dark.

It prefers a spot where the cliffs are not too precipitous, or where they exhibit sloping terraces grown with thrift and samphire, in which the Rock Pipit may find a suitable little cave for its nest, with a beetling brow in the shape of an overhanging piece of rock to protect it from the rain. There it will make its nest of grass, hair-lined, and deposit in it the five pretty green-grey eggs with evenly distributed reddish-brown specks. I have often sat on Cornish rocks and watched the Rock Pipit on the shore below, running along the lines of washed-in weeds, and evidently picking out small mollusks and shore-hoppers; I have found its nest also in the hollows of steep cliffs difficult of access.

The Chough (*Pyrrhocorax graculus*) was at one time a common bird in England, but it is now restricted to Ireland, the Isle of Man, parts of Wales, and south-west England. Cornwall was formerly regarded as its headquarters, and it was variously known as the Cornish Chough, Cornish Daw, Cornwall Kae, Market-jew Crow, as well as by other names not connected particularly with the Duchy; but so great have been the onslaughts upon it that the Cornwall County Council has had to get the Home Secretary to declare it a protected species, with a price upon the head of the miscreant who dares to take its eggs in the Western Duchy. Its plumage is black, with purple and green reflections, and its bill and legs bright red. It fortunately nests in difficult places in high cliffs, where it makes the nest which Yarrell describes as built of "sticks lined with wool and hair," in which it lays "four or five eggs of a yellowish-white colour, spotted with ash-grey and light brown."

The Jackdaw (*Corvus monedula*) often builds in holes in high cliffs. We have found its nest far inside a rabbit hole that was probably never intended by the rabbit as a means of entrance to or exit from his burrow, but as a secluded place whence he could look out upon the blue sea hundreds of feet below. But the rabbit had probably been evicted, or had fallen a prey to the ravens that built hard by, and the Jackdaw had taken possession. I knew the nest was there from watching the excursions of the old birds, but it was only by lying along a dangerously narrow ledge and pushing my arm in, right up to the shoulder, that I could feel the nest and count the heads of the five young Jacks. The nest of the Raven (*Corvus corax*) was in a hole so high up the perfectly straight face of the rock, that its entrance could only be reached by a person swung from the cliff fifty feet above it. All one could do was to watch the young birds fly out in a batch and hear the parental croaking that was evidently intended as approbation of their progress.

But to get to the distinctly maritime species, and first those of the Pelican family. We have two native species of Cormorant, the Common Cormorant (*Phalacrocorax carbo*) and the Shag (*P. graculus*). The Common Cormorant, Great Black Cormorant, Cole Goose, or Skart, as it is variously styled in different localities, is a bird of the rock-bound

SHAG.

coast, where there are detached masses of rock forming little islets, and where the face of the high cliffs is broken into narrow ledges. Such a coast will have at distances of a few miles its Shag-rocks and Shag-stones, which are well-marked by an abundant coating of white-wash. These are the resting-places whither the Cormorant and Shag repair to eat and digest the fish they have just captured, two or three miles further along the coast it may be. Similarly you may always tell the situation of their nests on the high ledges of the cliffs, though from the shore no part of the nest itself may be seen: but the streaks of white-wash splashed far down the precipice have only to be followed upward with the eye, and they will be seen to end just below a narrow shelf. Upon that shelf the rough nest is placed, and in it lie the four or five bluish-green eggs that afterwards become coated with a white crust. The colouring of the adult is more or less green, with patches of white on the neck and the outer part of the thighs. In winter these white patches become less conspicuous, and the green of its upper parts changes to a rusty black.

The Shag, Green Cormorant, Crested Cormorant, or Crested Shag, may be distinguished from the other species by its entire green colour and its smaller size. This difference in measurements, etc., may be put into a readily-seen form thus:—

	BILL	WING	TAIL FEATHERS	EGGS
P. carbo.	4½ to 5 ins.	14½ ins.	14	4 to 6
P. graculus	3½ „	10 „	12	3 „ 5

The two species are very similar in their habits, watching for fish from their favourite rock, and when their prey is seen diving after it. Sometimes they skim the waters and suddenly dive in after fish seen through the water.

Our other British Pelican is the Gannet or Solan Goose (*Sula bassana*), whose breeding-places are restricted to certain islets off the north-western coasts, such as the Bass Rock, Ailsa Craig, and some of the jutting headlands of the Hebrides, where they build extensive nests of dry grass, seaweed, and anything else that happens to be handy when they are building. On this they lay their solitary white egg, which must be described as small in proportion to the bird, and ridiculously small compared with the nest. Great numbers gather at their favourite breeding-places in the spring, and they are then said to be very tame. In autumn these great

assemblages, with the new generation they have reared, break up, and the individuals distribute themselves widely over the seas, where they follow at a height the shoals of fish, suddenly diving straight to the mark and capturing the fish selected for their prey.

SOLAN GOOSE.

The Common Grey Heron, or Hern (*Ardea cinerea*), though not usually reckoned among maritime birds, is frequent on our shores, and may often be seen to fly along, then settle with his feet in the water beside some grey rock where he is all but invisible, and watch for fish and other marine creatures.

The ill-named Oyster Catcher (*Hæmatopus ostralegus*) is well distributed along our shores, and it is no uncommon thing to hear its loud and shrill rattling pipes, and turning at the sound, to see its elegant form perched on a rock that is surrounded by water. In such a prominent position its black

OYSTER CATCHER.

and white plumage, its red legs, and long red beak render it very conspicuous. I do not think it catches many oysters, unless they be the fragile Saddle-oysters (*Anomia*) from the rocks. The oyster of commerce and gastronomy (*Ostrea*) has too thick and large a shell for it, though its bill is a strong one and capable of breaking into the strongholds of small cockles, mussels, and Venus shells. So far as the coast is concerned it selects sandy shores for its breeding-places, where it may be seen running quickly up and down at the very edge of the water. It makes no nest, merely selecting the slight shelter of a tuft of grass on the higher, drier part of the sands, and there it lays its three or four clay-coloured eggs, spotted, blotched, and streaked with dark-brown, and arranged with their narrow ends close together. If there are four eggs in the clutch they will be arranged cross-wise.

The Purple Sandpiper (*Tringa striata*) must not be looked for by the summer visitor, for it only comes to these shores when it has donned its winter dress. The same may be said of the Knot (*Tringa canutus*), which is sometimes confused with the Purple Sandpiper, but they may be readily distinguished in winter dress by examining the upper tail-coverts. In *T. striata* these are quite black, but in *T. canutus* they are white barred with black.

The Sanderling (*Calidris arenaria*) is plentiful in spring and autumn, on the wet sands and adjacent rocks; its whitish underside showing distinctly. The Redshank (*Totanus calidris*) is also common on many of our shores; its winter plumage is uniformly grey above, white beneath; but in spring this changes to brown, spotted and barred with black on the upper parts, and the white of the lower surfaces becomes greatly modified by the many streaks and spots of dark brown.

The Curlew (*Numenius arquata*) in its seasonal migrations has usually some representatives upon the coast, though it is in the winter that they are most commonly seen there, especially in the south, where they spend the winter in flocks. The Whimbrel (*Numenius phæopus*), which is a smaller Curlew, is often found on the same shores in winter and spring. It may be distinguished not merely by the smaller stature, but by a difference in the colour and markings of the head. In the Curlew this has a light brown crown streaked with black, whereas in the Whimbrel the same part is dark-brown, with a pale buff stripe dividing the brown into two equal portions.

The Terns (*Sterna*) have been well-named Sea-swallows, and a flock of them flying, wheeling, and doubling, over the waters, presents a very close resemblance of movement to that of the real Swallow. Several species haunt our shores, some coming here to breed. Among these may be noted the Sandwich Tern (*Sterna cantiaca*), the largest of the genus that breeds here, which has a yellow-tipped black bill; the Common Tern (*S. fluviatilis*), which has a *red* bill, the tip of the upper mandible only being black, and that not a very good black; and the Little Tern (*S. minuta*) with a *yellow* bill, black-tipped, and a white forehead.

The Gulls (*Larus*) are, of course, abundant, and much time may be pleasantly spent sitting near the edge of a grassy cliff, or some distance up its face, and watching the flight of the gulls below, sometimes rapidly as though time and tide wait for no gull; at other times with an easy undulating motion as though it were not necessary to hurry about anything, and scarcely necessary to move a wing when sailing on a pair outstretched is so easy a matter. But the visitor is often puzzled to make out the difference between those he commonly sees, and this is never an easy task to a naturalist until he has spent much time with them and made himself acquainted with the colour changes of the birds from youth to adult age.

Below I am attempting to tabulate the most striking differences between several species, taking in each case the adult plumage.

	HEAD	BILL	PRIMARIES	BACK	LEGS	FEET
Black-headed Gull	Dark brown	Red	White, the fore-edge of the first black; hinder edges and tips of all black.	Pearly grey	Red	Red
Common Gull	White	Yellow, dusky at base	Black, spotted with white near tips.	Bluish grey	Lead-grey	With well-develop'd hind toe
Herring Gull	White	Yellow	Black, tipped with white	Ash grey	Flesh coloured	Pale flesh coloured
Lesser Black-backed	White	Yellow	Ditto	Dark grey	Yellow	Yellow
Gr'ter Black-backed	Pure white	Yellow	Black, tipped and barred with white	Pure black	Flesh coloured	Flesh coloured
Kittiwake Gull	Pure white	Dull clouded yellow	1st outer web black, others pearl grey tipped with black; tips of 4th and 5th spotted with white	Pearl grey	Dark lead coloured	Hind toe undeveloped

The Razorbill and the Guillemot are common birds on most of our coasts where there are cliffs, but we shall see them chiefly as swimming and diving birds as we walk along the shore. The Razorbill (*Alca torda*) when swimming carries its

RAZORBILL.

tail parts higher out of the water than the Guillemot (*Uria troile*), and is further distinguished by the high compressed bill with white transverse stripes, the white stripe from the bill to the eye, and the dark-brown throat. The Guillemot has a long, straight, pointed beak, white throat crossed by a greyish cravat, continued from the mottled black and white of the back of the head and neck. It is too common as a dead, sodden-plumaged bird in the rock-pools after winter storms, which prevent it fishing, and starve it to death. The legs and feet are greyish, the webs black.

The Black Guillemot (*Uria grylle*) breeds on cliffs in Scotland, Ireland, and Man, but in winter also visits the south and south-west coasts. Its summer dress is wholly black, save for a patch of white on the coverts, but in winter the black is all replaced by white and very pale grey. The legs and feet differ, too, from those of *U. troile* in being vermilion in the present species.

The Puffin (*Fratercula arctica*) is identified readily, wherever seen, by its conspicuous compressed orange beak of great depth from top to bottom. This gives it a humorous aspect that belongs to itself alone; but it is useful to it also, for it makes a very efficient cracking instrument wherewith

PUFFIN.

certain of the thinner shelled bivalves may be utilised for the Puffin's food. It is a great diver, and sometimes the habit is its ruin. I have a fine specimen that was drowned by running its head into the mesh of a mackerel-net, and failing to extricate itself in time to prevent death by drowning. Young specimens are sometimes blown in exhausted during winter gales. Many other birds are similarly overcome.

The pretty little Storm Petrel, or Mother Carey's Chicken (*Procellaria pelagica*), whose stuffed body is before me as I write, was blown in early in November, 1895. I tried to restore it to vigour, but it was too far exhausted to take food, and this appears to be the common condition of those that are blown in. On the same day many Gulls, Guillemots, and Shags were washing into our "porth," and several of these were cared for, restored to health, and given their liberty a few days later.

The Great Northern Diver (*Colymbus glacialis*) and the Fulmar (*Fulmarus glacialis*) are also winter visitants to most of our shores. It is thought the Diver may breed on some of our extreme northern islands, but there appears to be no evidence that it does so. It is a regular visitor to the Cornish coasts in winter, and it is well worth watching from some rocky headland. It is large and powerful, and excels not merely as a proficient diver with plenty of "staying power," but is a vigorous swimmer, and a very capable flier. It is a pity those who see it are not more content with the sight, instead of being possessed with the desire to get a gun and shoot it. One would like to see it more often alive, and less frequently adorning the halls of country houses near the coast.

The Fulmar is not of such general occurrence as the Diver, except in the far north—St. Kilda, Orkney and Shetland. St. Kilda is its breeding-place, and they are merely stragglers that put in an appearance during winter on more southern shores. The hooked-bill and tubular nostrils distinguish it from the gulls at a glance.

The Manx Shearwater (*Puffinus anglorum*) breeds on islands all down the western coast as far south as Scilly; it is therefore a more frequent visitor to our southern and western coasts, especially before and after it is engaged on the important work of hatching and rearing its solitary chick.

CHAPTER XIX.

SEAWEEDS.

IT is to the rocky shore we must first turn our steps, if we desire to obtain a wide acquaintance with the British Sea-weeds: that is the grand hunting-ground for the Phycologist. In the rock-pools he will find very many of the smaller species, and thickly coating the fringing rocks are the larger, tough and leathery species of *Fucus* and *Laminaria*, forming at once a breakwater that largely destroys the force of heavy seas, and a splendid cover for the soft-bodied creatures that swarm on the rock-surface, and feed on the plants that protect them from the fury of the waves. The ancients called them *inutiles algæ*, but in the ocean's ceaseless warfare with the land, the greatest obstacle the former has to encounter is the network shield of seaweed, that breaks the force of its heaviest blows. This is an utilitarian characteristic of the seaweeds, for which Britons, at least, should be thankful, quite apart from their minor importance as sources of food, physic, fodder, and manure, and their æsthetic qualities.

The whole class of Seaweeds, with the solitary exception of the Grass-wrack (*Zostera maritima*) belong to the flowerless division of plant-life, and to that section called Algæ. They are plants of simple organization, being innocent of wood or other complicated tissue; the whole plant being made up of cells, though in the higher families there is an approach to the formation of vessels and tissues. They are absolutely without roots, though the larger species are attached to rocks or other algæ, by what appears to be a root. This organ, however, does not penetrate into the substance to which it is attached, but is a mere sucker, sticking tightly to the surface, and taking no part in the absorption of food for the plant, which is effected by the entire general surface of the frond from the surrounding waters, in which it floats in a more or less erect attitude, but yielding to every movement of the sea.

The seaweeds are all reproduced by spores, but the structure and production of these differ in the different groups, some being asexual, and others the result of a distinct sexual process. To this matter we shall give further attention by and by.

The most striking seaweeds owing to their size and abundance are those comprised in the *Fucaceæ*, all the species of which are olive-brown in colour. There are four species of Fucus which are very abundant on our shores, as well as representatives of other genera included in the family. On the highest of the rocks, that only become covered by high spring tides,

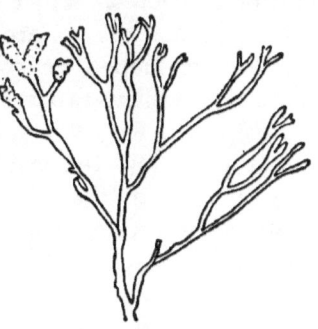

CHANNELLED WRACK.

and are only washed by rough seas, there grows in abundance a little leathery plant called the Channelled Wrack (*Fucus canaliculatus*). During the greater part of its existence it is dry and shrivelled and of black hue, but when covered by the tide it absorbs the water, and becomes soft, with an olive-brown tint. The frond is much branched, the segments long and narrow, the edges being turned in so that on one side they appear to be deeply channelled. It is the smallest of our species of *Fucus*, and is readily distinguished from its congeners, not only by its size, but by the channel-like folding, the absence of air bladders, and the lack of a mid-rib. At the extremity of some of its fronds there are irregular warty pod-like organs of a dark orange tint. These are known as the *receptacles*, and they contain the elements necessary for the production of spores, whereby the plant is reproduced. In one plant these elements will be all male (*antherids*), in another all female (*oogones*). If you will examine one of these club-shaped orange organs with your pocket-lens, you will observe that its surface is pitted with a considerable number of round pores, and if you cut across the whole body just on the edge of one of these pores, you will find it communicates with a globular cell in the substance of the receptacle. These cells are known as *conceptacles*, and their number corresponds to that of the pores. Their walls are clothed with a felt-work of threads, upon which are borne, in the male conceptacles, minute egg-shaped cells (*antherids*), which ultimately burst, and set free thirty-two or sixty-four tadpole-like bodies (*antherozoids*), each with two tail-like threads (*cilia*) attached to the under part. By the lashing of these organs they make their way out through the pore of the receptacle into the sea.

With the development of the antherozoids, a similar activity has taken place in the female conceptacles, where bodies approaching more to an ellipsoidal or spherical form (*oogones*) have appeared, and their contents have broken up into two, four, or eight smaller bodies (the *oospheres*). On their escape into the water, they are each surrounded by a number of the antherozoids, which pierce the substance of the oosphere, become absorbed in it, and so fertilise it. Development then commences in the oosphere, and it gives rise to a new *Fucus* plant. This form of reproduction is by no means common to the whole class of seaweeds; on the contrary, there are many important variations of it, which for want of space we shall be unable to refer to in detail. This is the highest type of reproduction in the Algæ.

The Channelled Wrack never exceeds a few inches in length, but another species, which agrees with it to the extent of possessing no mid-rib, varies from two to six feet. This is the Knotted Wrack (*Fucus nodosus*), which may be at once

identified by the possession of solitary bladders in the centre of
its rib-less frond, and producing a very gouty appearance at
intervals. These are air-cells, sometimes measuring two inches,
which give buoyancy to the plant. Above the bladders the
frond divides, and from these branches (but not at their
extremities, as in the Channelled Wrack) the pear-shaped
reproductive organs are produced.

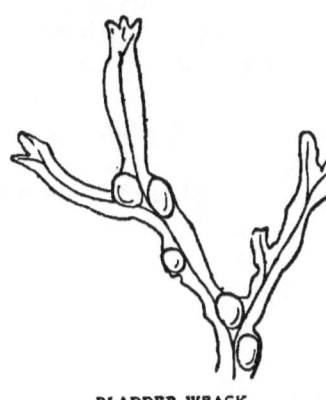

Another Wrack that possesses
these vesicles, is the so-called
Bladder Wrack or Black Tang
(*Fucus vesiculosus*), though there
is little danger of confusing the
two species. The Bladder
Wrack has a much broader,
flatter frond than the Knotted
Wrack, and a very distinct mid-
rib. The bladders, too, are
smaller, and instead of being
solitary, are arranged in groups
on each side of the mid-rib.
The plant is about two feet

BLADDER WRACK.

in length, and exceedingly plentiful.

Very similar, and equally
plentiful, is the Saw-edged
Wrack (*Fucus serratus*), with
flat, branched fronds and
mid-rib, the branches much
broader than in Bladder
Wrack, and the edges cut
into bold, sharp, distant
teeth. Its usual length is
from two to three feet. but
it may occur as long as five
or six feet. The width of
frond also varies, for it may

SAW EDGED WRACK.

be anything between half an inch and two inches. Where the frond branches the mid-rib becomes thicker and bolder. It is quite innocent of bladders. The name of the genus is founded upon the word *Phukos*, which is the Greek name for a seaweed.

Almost equally plentiful with those species of *Fucus* we have named, is the Pod-weed (*Halidrys siliquosa*), with long atten- uated compressed fronds, four or five feet in length, much branched, most of the branches being exceeding short, but others ending in air vessels. These are ribbed transversely, and bear a very close likeness to the seed-pods of the furze. They run out to a narrow point at the free end, and are divided into small air-chambers within. But there are other pods that contain the reproductive elements, and these may be known by the pores by which their surfaces are perforated. The name *Halidrys*, signifies sea-oak (Greek, *Hals*, the sea, and *drus*, oak), but the why and wherefore of the name are not easily determined. If the abun- dant pod-like vessels are kept in mind, there is no difficulty in knowing this species the first time it is seen.

POD-WEED.

At low-water, you will often find, at- tached to the rocks, a shallow horny cup, or button, of olive hue, about the size of a penny. This is the Sea-thongs (*Himanthalia lorea*), which gets its name from a very long, branched, strap-like growth from the centre of the cup. The cup is the frond—the plant proper—and the extraordinary straps, which may be half an inch wide and twenty feet long, are merely the receptacles containing the reproductive organs, which open by pores all over their surfaces. The receptacles are not produced until the second year of the plant's life, so that many examples will be met consisting of the cup-like frond only. It is a local plant, and not therefore to be found on all parts of the coast.

In the lower series of tide-pools, a tufted weed attracts the sight by reason of its brilliant iridescence, which often causes it to be plucked from its native pool, only to be thrown back again, for on emergence from the water all the beautiful play of colour has gone. It does not appear to have any common name, but to give it a chance of being popularly known, let us call it the Rainbow Bladder-weed (*Cystoseira ericoides*). The many branches of its frond are full of little bladders, whence its scientific name (*Kystos*, a bladder; *seira*, a cord), and it gets its specific title of *ericoides* from its habit somewhat resembling that of the Heath-plant (*Erica*). A tuft pulled up and carefully overhauled will afford the zoologist a number of diverse forms of life. Several species of crustacea make it their home, and the leaf-worms hide themselves in the centre of the little bush. Mollusks, sponges, and ascidians are there also, and the description of the animal inhabitants of such a tuft would make a fair chapter.

All the species of seaweeds to which we have already referred, are members of the class Fucaceæ. We have now to take a glance at other brown and olive weeds, some of which are the giants of the tribe, but which belong properly to the deeper waters, though every gale will make us well acquainted with their forms heaped up upon the shore. In this class—known to botanists as the Phæosporeæ—the reproduction is generally of a lower type than in those we have been considering. In the majority of forms there is no sexual process, the species being reproduced, as a rule, by zoospores, which are somewhat similar to the antherozoids of *Fucus*. They are produced in special cells, the contents of which break up into a number of these zoospores, which escape through a pore, and germinate.

Getting down into a drang at extreme low spring-tide, we shall find the rocks to seaward covered with Tangle (*Laminaria digitata*), whose huge round stems clasp the rocks with their claw-like false-roots. The leafy portion is broad, of a pale olive-brown, and slit up into several sections, so that the whole frond has a rough resemblance to the diverging fingers of a huge hand : hence its name, *digitata* (having fingers). The substance of the frond is thick and leathery. A species with undivided glossy narrower fronds, puckered and frilled, is the Sugar Tangle (*Laminaria saccharina*), so-called because, when drying, it produces on its surface a white powder of a sweet taste, called *mannite*, or manna. This substance can also be obtained from the cells by maceration; Subsequent evaporation of the brew results in a deposit of crystals. This is the species that inland trippers carry away on their visit to the coast to act as a hygrometer, hanging it on a nail, and feeling it from time to time to find if it is dry and hard, or moist and pliable, for its cells readily absorb moisture from the atmosphere, and as readily part with it when the air is again dry and clear.

A third species is called the Sea Furbelows (*L. bulbosa*), and it may often be found washed up in great heaps after a storm. It springs from a great hollow sphere, which is perforated, and thus affords a home for many creatures. This so-called bulb is sometimes a foot across, and from its stem there is a great expanse of thin leather split up into many broad ribbons. These three species, with the larger *Fuci*, are largely used by farmers near rocky coasts for manuring their fields, and in former days, more widely than now, they were employed in the manufacture of " kelp " and iodine. These Laminarians have the curious habit of casting off the *lamina* or blade of the frond each year, by a constriction above the stem, whence a new one grows. This, too, it should be stated, is the growing point, the blade increasing in length by additions near the stem, instead of by the lengthening of the free end. The spores are produced in large patches upon the surface of the frond.

The Badderlocks or Murlins (*Alaria esculenta*) of our northern coasts, belongs to this group, but is distinguished from the *Laminaria* by the possession of a mid-rib or central nerve.

The stem is short and cylindrical, and the blade of the frond ranges from three to twenty feet in length, usually much torn by the waves. · There are a number of finger-like receptacles given off by the stem, and in the outer coats of these are the conceptacles bearing the spores. The plant is used as food by the poorer classes resident on the shores where it is plentiful, and is eaten raw, when it is said to be the best of our esculent seaweeds; the parts preferred are the mid-rib and the receptacles. "Badderlocks" is a corruption of Balder's locks, the split fronds being likened to the locks of the Scandinavian hero Balder, to whom all plants except mistletoe swore fealty.

To this class also belongs the slender and very extensive Sea Lace (*Chorda filum*), which consists of a rounded frond, hollow, and without branches. It is remarkable how tenacious the thong-like, slimy fronds are, and it is not difficult to imagine the difficulties of a swimmer who should have to force his way through a bed of them. The tubular interior is divided up into a number of cells by transverse partitions; and the spores are embedded in the outer surface. It prefers a sandy or muddy bottom in creeks and harbours, and in such places it grows in dense patches, the fronds attaining a length of from twenty to forty feet. The free end is constantly dying off, but the plant increases by growth at the lower end, just above the false roots.

The Fennel-leaved Net-weed (*Dictyosiphon fœniculaceus*) is abundant in rock-pools all round our coast. As its name implies, the frond is much branched and thread-like. It is a light olive in colour, and grows in tufts on stones and larger weeds. The arrangement of the cells in the walls of the frond produce a net-like appearance.

Everyone knows the thin flat transparent fronds of Sea Lettuce (*Ulva latissima*), which grows everywhere on the coasts, its margins crisped, folded, torn, or otherwise diversified by Nature, or the many things that feed upon it. It is mentioned out of its place here, in order that we may bring into its proper order a plant that is frequently taken as a mere aberration of the *Ulva*. This weed is the *Asperococcus turneri*, a hollow green bladder on a short stalk, and rough with the spore-bearing organs. It is commonly found adhering to stones between tide-marks.

PEACOCK'S TAIL.

One of the most beautiful of our sea-weeds is known as the Peacock's-tail (*Padina pavonia*). It is really a tropical species, but its range of distribution extends to our most southern shores, and, strange to say, without suffering any deterioration in its brilliance of hue or its stature. From a very narrow base the frond gradually expands to a broad fan-shape, and the edges are curled in so that it assumes a cup-shape. But the chief beauty of the plant is given by a number of concentric lines and bands. Several of these bands are white, as though they had been chalked: their colour is in fact due to a chalky powder, *calcium carbonate*, which is secreted by the plant. Many of the lines are formed by a fringe of glistening hairs, which reflect the light and break it up into all the colours of the spectrum, and a more distinct fringe decorates the upper margin of the frond. Reproduction takes place by the formation of large spores, which are found in heaps between the zones. These are known as *tetraspores*, because the contents break up into four smaller spores. Sometimes this weed is what botanists term *proliferous*, that is, it produces new plants upon its frond. Like all those showing iridescence, it is a much more beautiful species in the water than in the herbarium; though it is not without beauty there, and it is a prize eagerly sought by collectors.

It is worthy of note that freshwater Algæ are, with very few exceptions, green, whilst few of the marine species are truly green; brown and olive, and red, are the prevailing hues. The green marine weeds are nearly all found in shallow water. Of course, they all possess the green colouring matter called *chlorophyll*, but in the deep-water species, according to Murray and Bennet, "it appears to be essential . . . . that the green colour of the chlorophyll should be masked by a coloured pigment, red in the case of the Florideæ, brown in those of the Phæosporeæ and Fucaceæ." It is from these latter classes our examples have been already drawn; we must now give a turn to the Florideæ, which contains many of the most popularly sought species, because they are often so charmingly tinted and so delicate in structure.

It must not be supposed from the foregoing remarks that the whole of this class are red weeds; the majority are not only red but brilliant red; whilst others are purple, brown, yellowish, or dirty-white. They are chiefly small weeds, but they make up for the want of stature in their delicacy of texture and fineness of division.

A very beautiful genus of delicate red and purple weeds, chiefly growing upon the larger and coarser kinds, is called *Callithamnion* (Greek, *Kalli*, beautiful, and *thamnos*, shrub). Some attain the length of half a foot, but most of them are much smaller. They require careful examination with lens or microscope to decide the species, and oftentimes in order to distinguish them from other finely branched red-weeds. For their proper discrimination we advise reference to a book devoted exclusively to Seaweeds, such as Lands-

borough's, Gray's, or the splendid " Phycologia Britannica "
of Harvey. The general characters of the genus are: frond
branched, often pinnate, consisting of jointed threads, with
tetraspores scattered along the branches.

Other small red seaweeds will be found, representing several
genera, but they require the assistance of coloured figures to
make descriptions interesting and useful. There is the silky
Ptilota, with finely divided fronds, consisting of cells alternately filled with a pink and a transparent fluid; the rosy red
Griffithsia, with thread-like fronds and clear transparent joints;
the forking threads of *Ceramium*, their tips curled in towards
each other; the exquisite *Plocamium*, with its flat crimson,
hair-like branches, toothed on one side only.

The Coralline (*Corallina officinalis*) grows in every pool, and
its stony-coated joints are well-known, though it is a shock
to some persons to find it classed among plants, when they
had long imagined it to be related to the corals of which neck-
laces and islands are constructed. There are, in fact, several
genera whose members secrete carbonate of lime, and so hide
their vegetable character. The coralline was, however, once
soft and flexible. *Melobesia* is equally stony, but grows in thin
horizontal pink and purple plates or solid masses. A little
weak muriatic acid will soon dissolve the lime, and reveal its
true character. *Jania* somewhat resembles *Corallina*, but its
branches are exceedingly slender, and much shorter than
Corallina.

Among the larger red weeds that will attract attention at
low-water, is the coarse textured *Halymenia ligulata*, of dark
crimson hue, whose strap-shaped fronds support other straps
by very slender attachments. It is closely related to *Rhodymenia palmata*, a very common red seaweed, that is eaten in
Scotland, Ireland, and on the West coast of England, under
the name of Dulse or Dillisk, though it is said to be a not very
desirable food when anything else is to be obtained. Its
fronds are roughly fan-shaped, consisting of a great number of

radiating ribbon-like lobes, of a purple colour. Its texture is like that of parchment. It will be found parasitic upon the stems of *Fucus* and *Laminaria*, at very low-water. A more slender and ragged, thin textured species is *Rhodymenia jubata*, with irregular outgrowths all along its edge, some of these fringes developing into long lobes. Another species that is also eaten as Dulse is the *Iridea edulis*, which glitters with bluish iridescence when immersed. It has fronds about six or seven inches long, expanding into a broad oval at the free end, and thinning off to a wedge-shape at the base. It is represented in the illustration of the Prawn, on page 163.

The Pepper Dulse (*Laurencia pinnatifida*) is a much smaller species, that grows abundantly in the pools and the rocks around them, standing the repeated scorching-up when the tide withdraws, as well as does the Channelled Fucus, its companion. It roughly resembles a miniature Polypody fern, but of a purple colour.

Another edible weed is the well-known Irish Moss or Carrageen (*Chondrus crispus*), which was in such favour years ago as an invalid's food. It is well shown in the illustration, but is subject to great variation, especially as regards colour, ranging from greenish-white, and yellow, to a dull purple. In some of its forms it closely resembles *Gigartina mamillosa*, to which it is not very distantly related, and the danger of confusing the two is increased by *Gigartina* often growing amongst *Chondrus*. The tips of *Gigartina's* frond, however, are usually broader than those of *Chondrus*, and the frond is rough, with little tubercles like

CHONDRUS CRISPUS.

grape-stones (Greek, *gigarton*), which contain the spores. The usually purplish fronds will be found, on cutting them across, to be not solid, as they appear, but composed of delicate threads, in a firm clear jelly.

A pretty little red weed, that is abundant in the rock-pools, growing upon other weeds, is the *Chylocladia parvula*, which has swollen, cactus-like ovate joints, of a clear red, appearing as though they were skins filled with liquid. It is allied to *Plocamium* and *Rhodymenia*.

The most striking of all these red-spored algæ, at least, so far as the British flora is concerned, is the (for a seaweed) extraordinary Ash-leaved Seaweed (*Wormskioldia sanguinea*), whose frond has a distinct leaf-like form, with a mid-rib and branching nervures. Its texture is so very thin, that in spite of its beautiful rosy tint, if a specimen were laid upon this page, the print could be read through it. Its margins are more lax than the mid-rib, so that when mounted for the herbarium, the edges show many foldings over. The plant was formerly placed in the genus *Delesseria*, but is now separated on account of important differences in the matter of propagation. In this species minute leaf-like organs spring from the mid-rib, and may be taken for young plants springing from the parent, but these are really the bodies that bear the spores.

ASH-LEAVED SEA-WEED.

The Winged Delesseria (*Delesseria alata*) is a finely and intricately branched plant, of a rich dark crimson colour, with a suggestion of a mid-rib, along each side of which is a narrow expanse of thin membrane, the "wings" of its popular and technical names. It occurs in thick tufts on the stems of *Laminaria digitata*.

My space is getting rapidly used up, though I have only been able to mention a few of our fairly common seaweeds. There are still two or three that I must mention. One of these is an exceedingly pretty little form, which would be very like a soft feather that has been cast by one of the greener varieties of the canary-bird, if it were not so vividly green. The weed is called *Bryopsis plumosa*. It will be found growing on the shaded walls of deep pools, and if the eye is placed · just over the edge of the pool, the Bryopsis will be found growing at right angles with the wall, and looking so very feathery that it will be identified at once.

Another green weed that should be mentioned is the *Enteromorpha compressa*, of the same texture as the Sea Lettuce (*Ulva*), already mentioned, but forming a narrow tube of rugged shape, that is ordinarily collapsed, but sometimes inflated with oxygen gas. It is represented in the illustration of the Sand Launce on page 275. A tuft growing on a stone or limpet-shell, is a valuable addition to the aquarium, for it will continue to grow, and many of the animal inhabitants will find their food in it. Crustaceans, fishes, and mollusks are all fond of it.

The reader who has patiently accompanied me thus far, will probably make up his mind to preserve some of these beautiful weeds, and I should strengthen any such intention; but let me advise that some care be expended upon the work. ·Select your specimens with care, and be not satisfied until you have, by patient seeking and overhauling, secured fairly perfect examples with, as far as possible, the fruiting organs. These must be carefully laid out, and gently pressed between sheets of absorbent paper, just as in the case of flowering plants. But it should be always remembered that the specimens as taken from the sea are more or less coated with salt, and will never thoroughly dry until this is removed. The first care then should be to well rinse them in clear soft water, a few specimens at a time, to avoid leaving any for long in the fresh water, which rapidly destroys certain species if they are left in it for more than half an hour. Lay them out in as natural a manner as possible, separating the delicate divisions of the frond with a camel-hair brush. When thoroughly dry and hard, mount specimens of one species only on the same sheet of paper, and neatly write the name of species near the bottom left-hand corner, and near the opposite margin, the place where, and the date when, collected.

CHAPTER XX.

FLOWERS OF THE SHORE AND CLIFFS.

JUST as in walking along the shore we have on one hand a region inhabited by specialised races of animal and plant life altogether different from those of the land, so also on the landward side we have flowering plants distinct in most cases from those found but a short distance inland. Strictly speaking, the stretch of shore, whether it be shingle, sand, or rocks, does not form a barrier separating sea plants from those of the land, for the terrestrial and the submarine overlap through the medium of the frondose lichen, *Lichina pygmæa*, which, belonging to a terrestrial group, spends half its day in the water and the other half exposed to the atmosphere. The pretty Sea-Milkwort (*Glaux maritima*) takes up the connecting thread on the land side, and establishes its roots and woody base jammed in the crevices of rocks, where they must absorb more salt water than fresh, and at times it must be entirely covered by the sea. That this salt is thoroughly congenial to its nature we may gather from the fact that the only inland localities where *Glaux* grows are the salt-producing districts. It attains to only a few inches in height, and its small, smooth, stalkless, glaucous leaves are thickened like many other shore plants, and dotted all over with minute pits. The flowers are devoid of petals, but the bell-shaped calyx is coloured of a flesh-tint, and sprinkled with very small dots of crimson. Its flowering period is from May to August.

In similar situations grows the beautiful little Sandwort Spurrey (*Spergularia rubra*), with many slender compressed, ruddy stems radiating from a woody rootstock; the leaves

U

slender, awl-shaped, unequal in size.   Petals fine, bright rosy ;
anthers yellow.   Flowers June to September.

Like conditions of life often produce similar effects on
different organisms.   Growing close to the Sea-Milkwort, just
above high-water mark, and continuing thence some distance
up the cliffs, is the Samphire (*Crithmum maritimum*), with
similar woody rootstock similarly wedged in rock crevices, and
with all its parts thickened.   The glaucous leaves are cut up
into cylindrical fleshy segments, and the yellow flowers are
borne in clusters, the fleshy stalks of the individual blossoms
radiating from a common centre like the ribs of an umbrella.
It may be unnecessary to explain that this type of flower-
cluster is characteristic of the Natural Order Umbelliferæ, to
which the Samphire belongs, and that it is to the same order
that such well-known plants as carrot, hogweed, fool's-parsley,
and celery belong.   Samphire is much sought for pickling,
and this has led to its extermination on many parts of the
coast.   It flowers from June to September.

Fennel (*Fœniculum officinale*) is another seaside umbellifer,
and its tall, straight, and polished stems may be found growing up the face of the cliffs, the much-divided feathery leaves
producing a green cloud-like effect.   The same glaucous tint
characterises the whole plant, except that the flowers are
yellow.   July and August are the months in which it may be
found in blossom.

One other umbelliferous plant that is strictly confined to the
shore is the so-called Sea-Holly (*Eryngium maritimum*), though
this must be sought not on the rocky cliffs, but on sandy
shores.   Its dense heads of pale-bluish flowers without a
stalk nestle close to the broad and spiny-edged glaucous
leaves (glaucous again), that bear a wonderful *primâ facie*
resemblance to those of the unrelated holly-tree.   It flowers in
July and August, but the plant is easily recognised out of its
flowering season by means of the bold leaves.

But the glory of cliff vegetation to my mind is the beautiful

Thrift or Sea-Pink (*Armeria maritima*), whose tufts of thick, narrow, grass-like leaves extend from the wave-washed rocks right up the cliff-side, and over the stony hedges at the top. It flowers sparingly all the year round—I have gathered it within a few days of Christmas—but the brilliant display is in April and May, when every clump supports many long-stalked, half-round heads of the rosy flowers, that make so beautiful a setting for the nests of the cliff-building birds. Thrift is not absolutely peculiar to the coast, for it is found also on high mountains; in the Scottish Highlands it occurs at an altitude of nearly four thousand feet above the sea. There is a larger and more rigid species (*A. plantaginea*) that grows on sandy banks in Jersey.

A relation of the Sea-Pink is the Sea-Lavender (*Statice limonium*), which grows where sand and mud are more abundant than rocks, and in some places covers the sand-hills with a growth not unlike that of the heather on inland sand-hills, and at a distance the purplish flowers are very suggestive of heather in such a situation. They are not gathered into a compact head as in Thrift, but are scattered along a branching spray. It has a creeping rootstock of a woody character, from which all the leaves spring directly. These are oval in general outline, running off to a point at the upper end. It flowers from July to November.

On the sandy shore where grows the Sea-Lavender there will, in all probability, also be seen a bold-leaved plant, with large, golden yellow flowers, which the tyro in botany will notice at a glance has some sort of relationship with the familiar Eschscholtzia of the garden. It is the Yellow Horned-Poppy (*Glaucium luteum*), and the above-mentioned tyro will say that this time the glaucous hue of the leaves (from which this species and *Glaux* both derive their scientific names) is not wholly due to its seaside habit, for the same hue is characteristic of Eschscholtzia and the Opium Poppy (*Papaver somniferum*), which are cultivated flowers. Quite so, but probably their original home may be near the sea, though the texture of their leaves is not so fleshy as in our maritime plants of glaucous hue. The bold, rough leaves make the plant conspicuous even in winter. The name of Horned Poppy is suggested by the form of the seed vessel, which is similar to that of Eschscholtzia, but thicker. It is a prominent feature of the flower—which loses its petals after one day's blossoming—but they ultimately extend to a foot in length. The flowers may be found from June till October.

Saltwort (*Salsola kali*) is also a plant of the sandy shore, with rigid brittle stems, striped and bristly, and fleshy, glaucous leaves, nearly cylindrical in shape, with spiny points. At their base the leaves become broader and partially clasp the stem. The little flowers are leafless, borne in the axils of the leaves, and to be seen only in July and August. This is one of several plants that were formerly burned to make Barilla, an impure carbonate of soda, much used in the manufacture of soap and glass, before the discovery of the cheap production of soda from common salt.

If my friend the reader is acquainted with the beautiful white-flowered Bladder Campion (*Silene cucubalus*), of inland hedgebanks, and he should chance to come upon the nearly allied Sea-Campion (*S. maritima*), he will think he has the old familiar plant, so closely are the two related; but a comparison will convince him there are differences. For instance, the stems of *maritima* are shorter and less erect than those of *cucubalus*. The flower cluster (*panicle*) is in *cucubalus* many-flowered; in *maritima* the flowers vary only between one and four in a cluster, and their petals are not so deeply cleft. The two scales that are obvious at the base of the broad part of the petal in *maritima*, are very obscure in *cucubalus*. *Maritima*, too, has smaller leaves and larger flowers, and the scales (*bracts*) below the flowers, which are dry and semi-transparent in *cucubalus*, are here more fleshy. It flowers from June to September.

Everybody is well acquainted with the pretty Field Convolvulus, or Small Bindweed (*Convolvulus arvensis*), and as they
have just seen it growing in abundance in the fields they
passed through on the way to the shore, they may reasonably
conclude that these larger, more richly-tinted blossoms that
grow on the sandy shore, are simply more luxuriant examples
of the same species. In reality they are produced by a distinct
kind, the Sea-Convolvulus (*C. soldanella*), which differs from
the common kind in the fact that clasping the base of the
flower and covering the sepals, there are two large leaf-like
bracts, whereas in *arvensis* these are small and placed at some
distance below the sepals. The leaves are fleshy, broader than
long, the stems are shorter, seldom more than a foot in length,
and very rarely do they twine around anything. The flowers,
as we have said, are larger and more richly coloured, only one
on a stalk, whilst the common sort have usually from two to
four.

The Sea-Rocket (*Cakile maritima*) is abundant on most
sandy shores. It is a large succulent plant, about two feet in
height, with zigzag branches, and smooth, fleshy, glaucous
leaves; flowers with four purplish white petals, arranged
cross-wise. The flowers are succeeded by large succulent
pods, that are divided into two by a cross-partition; each
chamber contains a solitary seed. It is this pod that is most
· likely to arrest attention. It flowers in June and July.

Wall Pennywort, or Navelwort (*Cotyledon umbilicus*), is an
abundant weed in the rocks and walls of the west coast, but
travels no further east than to Kent. Its tuberous rootstock
is wedged into the crevices of the rocks and cliffs, or between
the flakes of which stone dykes are built. The leaf, as the
name suggests, is round, with the stalk in the centre; it is
also thick and fleshy, the severity of the margin taken off by a
series of low, rounded teeth. Some of these leaves are large—
as much as three inches across. When the flowering stalk
makes its appearance, another type of leaf comes with it—

spoon-shaped. The flower-stalks bear drooping cylindrical flowers, greenish-white in hue, densely crowded, and all hanging downwards. It is a very striking ornament of the places where it is common, especially from June to August, when it flowers.

In company with the Navelwort, on rocks and walls, will be found one, if not two, species of Stonecrop (*Sedum*). One of these, the common Yellow Stonecrop or Wall-Pepper (*Sedum acre*), is too well-known to need describing. The other is the White Stonecrop (*Sedum anglicum*), of similar habit, but with the inevitable glaucous leaves (those of *S. acre* are *not* glaucous, but bright green); though sometimes these take on a reddish hue. The flowers are more star-like than those of *S. acre*, and of a whitish or pinkish colour—in evidence from May to August.

I do not pretend to furnish an exhaustive list of the plants of the sea-shore: that properly treated would make a volume by itself. Such as I have mentioned belong almost solely to a habitat where they can receive the salt spray upon their leaves. Mention should also be made of the Sea-Spleen-wort (*Asplenium marinum*), among ferns, that loves to grow over the entrance to a sea-cave, there hanging down its boldly-cut and well-varnished dark-green fronds, well out of reach. Then there is a distinctly marine *Carex*, the Sea-Sedge (*Carex arenaria*), which shares with Marram-grass (*Ammophila arenaria*), the work of binding the sands together with its thick, creeping rootstock.

But the seaside visitor, with botanical tastes, will find the shores abundant in vegetation generally, and he must have recourse to a special handbook to help in their discrimination.

Were it not for fear of laying himself open to a charge of presumption, egoism, favouritism, and a few other isms, the author would, in this connection, recommend his own "Wayside and Woodland Blossoms," Second Series,* which includes many of the maritime flowers.

CLASSIFIED INDEX

TO

Species mentioned in the foregoing pages.

Balanus balanoides, 183;
B. porcatus, 184
Pyrgoma anglicum, 183

Sub-class Malacostraca—

Talitrus locusta, 173
Orchestia littorea, 173
Caprella linearis, 174
Corophium longicorne, 175
Ligea oceanica, 172
Idotea marina, 173
Campecopea hirsuta, 174
Næsa bidenta, 174
Mysis flexuosa, 169
Leander serratus, 160;
L. squilla, 165; L. fabricii, 165
Pandalus montagui, 166
Hippolyte varians, 166
Crangon vulgaris, 167
Astacus gammarus, 152
Palinurus vulgaris, 152
Upogebia stellata, 171
Callianassa subterranea, 170
Galathea squamifera, 150
G. nexa, 151; G. dispersa, 151; G. intermedia, 151; G. strigosa, 151
Porcellana platycheles, 147; P. longicornis, 149
Eupagurus bernhardus, 144; E. prideaux, 144
Maia squinado, 152
Gonoplax rhomboides, 156
Corystes cassivelaunus, 155
Portunus puber, 140
Carcinus mænas, 139
Pilumnus hirtellus, 138
Xantho incisus, 137; X. hydrophilus, 137
Cancer pagurus, 131

Sub-kingdom—**Mollusca.**

Class *GASTROPODA*—

Fissurella græca, 223
Emarginula reticulata, 223; E. rosea, 223
Patella vulgata, 207
Patella pellucida, 211; P. lævis, 211
Acmæa testudinalis, 211
Phasianella pullus, 221
Trochus cinereus, 221; T. magus, 222; T. zizyphinus, 221
Scalaria communis, 219
Janthina fragilis, 222
Cerithium reticulatum, 219
Turritella communis, 219
Natica monilifera, 216
Pileopsis (or Capulus) hungaricus, 223
Littorina littoralis, 220; L. littorea, 220; L. rudis, 220
Rissoa ulvæ, 219
Aporrhais pes-pelicani, 219
Cypræa europea, 215
Ovula patula, 216
Erato lævis, 216
Murex erinaceus, 214
Fusus antiquus, 214; F. contrarius, 214
Buccinum undatum, 213
Nassa incrassata, 213
Purpura lapillus, 212
Aplysia depilans, 229
Doris johnstoni, 227; D. tuberculata, 227
Eolis coronata, 227; E. papillosa, 228

Class *SCAPHOPODA*—

Dentalium entalis, 224; D. tarentinum, 224

INDEX.

(The *popular* names are printed in *italics*.)

SELECTIONS FROM

MESSRS. JARROLD & SONS' NEW BOOKS.

Messrs. Jarrold and Sons'

"GREENBACK" SERIES OF POPULAR NOVELS

BY AUTHORS OF THE DAY.

In crown 8vo, cloth, 3s. 6d. each.

The *Western Mercury* says :—" Success evidently attends the departure recently taken by that well-known firm of London publishers, Messrs. Jarrold and Sons, who deserve the thanks of the literary world for their efforts to reproduce in cheap, yet acceptable, editions, the works of favourite authors. Prominent among the books so treated are the novels of Miss Helen Mathers, which are being issued in the 'Greenback' series."

By HELEN MATHERS.

Cherry Ripe!
Story of a Sin
Eyre's Acquittal
My Lady Green Sleeves
Jock O'Hazelgreen
Found Out!
Murder or Man-
    slaughter?
The Lovely Malincourt

By CURTIS YORKE.

That Little Girl
Dudley
Hush!
Once!
A Romance of Modern
    London   [teau
The Brown Portman-
His Heart to Win
Darrell Chevasney
Between the Silences!
A Record of Discords
The Medlicotts
The Wild Ruthvens

By MRS. LEITH ADAMS.

Bonnie Kate
Louis Draycott
Geoffrey Stirling
The Peyton Romance
Madelon Lemoine
A Garrison Romance

By IZA DUFFUS HARDY.

A New Othello!

By SCOTT GRAHAM.

The Golden Milestone
A Bolt from the Blue

By T. W. SPEIGHT.

The Heart of a Mystery
In the Dead of Night

By MRS. H. MARTIN.

Lindsay's Girl

By E. M. DAVY.

A Prince of Como

LONDON: 10 AND 11, WARWICK LANE, E.C.

"GREENBACK" SERIES OF POPULAR NOVELS

—CONTINUED.

In crown 8vo, cloth, 3s. 6d. each.

By *ESME STUART.*

Harum Scarum

By *AGNES MARCHBANK.*

Ruth Farmer

By *J. S. FLETCHER.*

Old Lattimer's Legacy

By *MAJOR NORRIS PAUL.*

Eveline Wellwood

By
SOMERVILLE GIBNEY.

The Maid of London
Bridge

By *MRS. A. PHILLIPS.*

Man Proposes

By *FERGUS HUME.*

The Lone Inn   [Court
The Mystery of Landy
The Mystery of a Han-
som Cab
The Expedition of
Captain Flick

By *LE VOLEUR.*

By Order of the Brother-
hood !

By *JOHN SAUNDERS.*

A Noble Wife

By *MARGARET MOULE.*

The Thirteenth Brydain

By *MRS. E. NEWMAN*

The Last of the Haddons

By *EASTWOOD KIDSON.*

Allanson's Little Woman

By *ELEANOR HOLMES.*

Through Another Man's
Eyes

By *LINDA GARDINER.*

Mrs. Wylde

By *MRS. BAGOT HARTE.*

Wrongly Condemned

By *E. BOYD BAYLY.*

Jonathan Merle
Alfreda Holme   [ture
Zachary Brough's Ven-
Forestwyk

By *EVELYN
EVERETT-GREEN.*

St. Wynfrith and its
Inmates

By *MRS. HAYCRAFT.*

Gildas Haven

By *HUDE MYDDLETON.*

Phœbe Deacon

LONDON: 10 AND 11, WARWICK LANE, E.C.

New 6s. Novels.

Crown 8vo, Cloth, Art Linen, Gilt Top.

1. The Power of the Dog.
By ROWLAND GREY, Author of "In Sunny Switzerland," "By Virtue of His Office," "Lindenblumen," "Chris."

2. Black Diamonds.
(Authorised Edition). By MAURUS JÓKAI, Author of "Eyes like the Sea," "Dr. Dumany's Wife," "Midst the Wild Carpathians," "In Love with the Czarina." Translated into English by FRANCES A. GERARD, Author of "Some Irish Beauties," etc.

3. Judy a Jilt.
By MRS. CONNEY, Author of "A Lady House-breaker," "Pegg's Perversity," "Gold for Dross," "A Line of Her Own," etc.

4. Lady Jean's Son.
By SARAH TYTLER, Author of "Lady Jean's Vagaries." *[Shortly*

5. Colour Sergeant No. 1 Company.
By MRS. LEITH ADAMS, Author of "Bonnie Kate," "Louis Draycott," "The Peyton Romance," "Geoffrey Stirling," etc. *[Shortly*

6. The Inn by the Shore.
By FLORENCE WARDEN, Author of "Pretty Miss Smith," "A Prince of Darkness," "A House on the Marsh," "A Perfect Fool," etc. *[Shortly*

The Prisoner of Pekin, or the Swallow's Wing.
By CHARLES HANNANT. Graphically Illustrated by A. J. B. SALMON. *[Shortly*

JARROLD & SONS, 10 & 11, WARWICK LANE, LONDON, E.C.

"Fleur-de-Lys" Series of Illustrated 6s. Novels.

Crown 8vo, cloth elegant, bevelled boards.

When the Century was Young.

By M. M. BLAKE, Author of "The Siege of Norwich Castle," "Toddleben's Hero," etc., with numerous Illustrations by the Authoress.

In this story the hero and heroine are equally important. It is full of incident, gives glimpses of the Peninsula war, takes us through France in 1813-14, describes an English election of the period, while the interest culminates in the battle of Waterloo. The hero takes his part in the great fight, and is afterwards brought wounded to Brussels and laid in the street with some 50,000 companions ; an evil position from which he is bravely rescued by the heroine.

> *Black and White* says :—" It is written with artistic regard to detail. Miss Blake is among the few authors who can illustrate her own work. Many of her line drawings are extremely happy."
>
> The *Norwich Mercury* says :—" Miss Blake is in this new story just as vivid and as picturesque in her battle pieces as she was in her previous tale."
>
> The *Devon and Exeter Gazette* says :—" The book brings vividly to mind the conditions which existed in this country for some years before the battle of Waterloo, and the latter part is full of the martial element which filled the public mind of that period."

The Lord of Lowedale.

By R. D. CHETWODE, Author of "The Marble City," "The Fortune of Quittentuns." Illustrated by G. GRENVILLE MANTON.

> The *Morning* says :—" 'The Lord of Lowedale' is a tale of the sixteenth century, with plenty of hard fighting and desperate adventure. The saying that 'fortune favours the brave' is illustrated in the rousing story of the Count and his two valorous young friends."
>
> The *North British Advertiser* says :—" The book is one which will be read with avidity by youths. It is full, as stated, of hair-breadth escapes ; and though the great historical scenes then enacted in France are only incidentally alluded to, a fair reflex of the lawless state of the country at that period of its history is given. If the remaining volumes of the 'Fleur de Lys' series prove as interesting as this first one, they will assuredly form a very valuable addition to our book-shelves."
>
> The *Dublin Freeman's Journal* says :—" There are many worse and few better stories to while away the idle hours pleasantly, to visit old times and strange scenes, and make new friends in the shining fairyland of fiction."

JARROLD & SONS, 10 & 11, WARWICK LANE, LONDON, E.C.

The "Impressionist" Series.

Each in crown 8vo, handsome cloth, gilt, 3s. 6d.

Under this general heading Messrs. Jarrold and Sons propose to issue, from time to time, a volume of Short Stories. It is intended, so far as possible, to link story to story by some thread of mental connection, and, thus, to present to the reader a series of vivid and related pictures.

1. ## Dust in the Balance.
    By GEORGE KNIGHT. With Title Page and Cover Design by Laurence Houseman.

2. ## Some Women's Ways.
    By MARY ANGELA DICKENS. With Title Page and Cover Design by Laurence Houseman.

3. ## Blind Larry. Irish Idylls.
    By LEWIS MACNAMARA. With Title Page and Cover Design by Laurence Houseman.

The "Daffodil" Library.

A Series of Shorter Novels by Authors of the day, in convenient shape, and attractive binding.

Narrow 8vo, paper covers, 1s. 6d.; cloth, 2s.

1. ## The Jaws of Death.
    By GRANT ALLEN.

2. ## Sapphira of the Stage.
    By GEORGE KNIGHT.

3. ## The Kaffir Circus, and other Stories.
    By M. DONOVAN.

4. ## Because of the Child.
    By CURTIS YORKE.

JARROLD & SONS, 10 & 11, WARWICK LANE, LONDON, E.C.

"Flashes of Romance" Series.

Crown 8vo, Art Linen, 2s. 6d. each.

Blind Artist's Pictures, and other Stories.

By NORA VYNNE. 2nd Edition.

The *Spectator* says:—"These tales are remarkable less for their knowledge of life and the skill with which they sketch it, than for the freshness and originality of the idealism which is displayed in almost all of them."

The *Speaker* says:—"Miss Vynne exhibits a distinct talent for the short story. She writes with ease and spirit, and there is not only brightness, but a good deal of unforced feeling, in her sketches. In 'The Blind Artist's Pictures,' and its companion stories, she has produced a very readable little volume. Slight as the stories are, they are told with piquancy and point, whilst a pleasant vein of sentiment runs through the series."

Winter's Weekly says:—"All the stories are written in good, clear, simple language; the style is clever and epigrammatic, and the characterisation very natural and life-like."

The *Literary World* says:—"Has the charm of rare originality because the author has a sympathetic appreciation of character and a broad-minded realisation of the fact that human nature in all its phases, if naturally drawn, is interesting; and all the more interesting when the picture is taken from an unconventional point of view."

Wrecked at the Outset.

By THEO GIFT, Author of "Pretty Miss Bellew," "Victims," "Lil Lorimer," "Dishonoured," "An Island Princess," etc.

The *British Weekly* says:—"All three stories are vivid pictures of London life, and are written with great spirit and considerable artistic power."

The *Literary World* says:—"'Theo Gift' is realistic in the true sense of the word; she does not shirk a situation which confronts her, but there is no needless stirring up of social mud-puddles, and her work is artistic in its grasp of life and character."

Putt's Notions.

By MARIE HERVEY. New Edition.

The *Scotsman* says:—"The simplicity and tenderness of the author's style add force to the simple pathos of the tale. Those who like a simple love story, naturally told, will find more than one to their taste in this unpretending little volume."

The *Western Morning News* says:—"A certain charm of grace and beauty runs through the whole of them, and they display a power for dealing with the pathetic and tragic aspects of human life rarely met with in the sketchy fiction of the day."

The *Yorkshire Post* says:—"Mrs. Hervey has written five short stories—simple, unaffected, sensible, and interesting."

The *Western Daily Mercury* says:—"Each is written in the happiest of veins, and though, as we are assured, all are 'founded on facts,' they will prove none the less pleasing to the reader of whichever sex, and of whatever age, and may also exercise a beneficial influence on a large number."

JARROLD & SONS, 10 & 11, WARWICK LANE, LONDON, E.C.

Popular Holiday Books.

Poppyland.

Papers descriptive of Scenery on the East Coast. By CLEMENT SCOTT. Crown 8vo, cloth elegant, with 26 Illustrations by F. H. TOWNSHEND. Superior Edition, 2s. 6d. In attractive Paper Covers, 1s.

> "It is not surprising that a fourth edition of Mr. Scott's delightful little book is called for. It deals with the charming coast of the far south-east of our island, and the epithet 'Poppyland' given to that has a poetic ring about it that is very attractive. The vicinities chiefly dealt with are Overstrand and its vicinity, Cromer, Lowestoft, and Yarmouth."—*The Queen.*

Summer in Broadland ; or, Gipsying in Norfolk Waters.

Fourth Edition. Profusely Illustrated by the Gipsies. Crown 8vo, Illustrated Paper Covers, 1s.; or Superior Edition, cloth elegant, 2s. 6d.

> The *Queen* says :—"A hardy manual based upon actual experiences. It is nicely illustrated, and is sure to please and instruct visitors to the Broads of East Anglia."

> The *Scotsman* says :—"A delightful little volume, with many dainty illustrations, as a result of a holiday among the little-known Norfolk and Suffolk Broads. There are many capital descriptions of scenery, and the book is marked by bright and vigorous writing."

Sunrise-Land.

Rambles in Eastern England. By ANNIE BERLYN, Author of "Vera in Poppyland," etc. Illustrated by A. RACKHAM and M. M. BLAKE. 3s. 6d.

> "Full of warm sympathy for picturesqueness and colour, written in an agreeable fashion, and evidently compiled from an intimate knowledge of the country and its inhabitants."—*Pall Mall Gazette.*

Two Knapsacks in the Channel Islands.

By JASPER BRANTHWAITE and FRANK MACLEAN, B.A., Oxon. Illustrated by VICTOR PROUT. Crown 4to, Pictorial cover, 1s.

JARROLD & SONS, 10 & 11, WARWICK LANE, LONDON, E.C.

THE DICTIONARY OF BRITISH MUSICIANS. From the Earliest Times to the Present.

By FREDERICK J. CROWEST, Author of "The Great Tone Poets," "A Book of Musical Anecdote," "Phases of Musical England," "Advice to Singers," "Musical Groundwork," "Cherubini," &c.

Price One Shilling.

"It is the most complete list of British musicians yet published. So far as we have been able to test the volume, it is accuracy itself."—*Musical Standard.*

"A handy little book of reference, giving in a concise form a list and particulars of those musicians who have done service in aiding the cause of our national music from the earliest times to the present."—*Westminster Gazette.*

"'The Dictionary of British Musicians' gives in a concise form a list of native musicians, composers, organists, instrumentalists, and singers. If not a musical nation, we can, however, claim some 3,500 names, all entitled to be recorded in such a work."—*Star.*

"Mr. F. J. Crowest has managed to pack a good deal of information into a very small space."—*Globe.*

"It is a valuable addition to musical biography dictionaries, and will no doubt supply a want."—*St. Paul's.*

"A capital reference book, and no one interested in music should be without it."—*Illustrated Sporting and Dramatic News.*

"An admirable little compilation it is, clearly arranged and accurate, bringing the result of vast research and encyclopædic special knowledge to the hand of all."—*Baker's Record.*

"The author deserves the thanks of all who compose or find pleasure in music."—*Western Morning News.*

"This dictionary will be found of much value for the purposes of reference, while the long list of names will astonish many who think lightly of native talent."—*Eastern Daily Press.*

"This little volume cannot fail to be of service."—*Birmingham Daily Gazette.*

"Mr. Frederick J. Crowest (author of the 'Great Tone Poets') has rendered a distinct service by the publication of his very cheap and useful 'Dictionary of British Musicians,' in which not only dead, but living celebrities are succinctly dealt with."—*Western Mail.*

"A handy little volume, which meets a long-felt want."—*Belfast Evening Telegraph.*

"It contains the names of authors and writers upon music, which, so far as we are aware, have not hitherto been found in any dictionary. It claims to be a dictionary from the earliest times to the present, and is quite up-to-date."—*Sheffield and Rotherham Independent.*

JARROLD & SONS, 10 & 11, WARWICK LANE, LONDON, E.C.